Marcos Antonio P. Coelho
Pollylian A. Madeira
Rivelino P. Coelho

Learning theories

Marcos Antonio P. Coelho
Pollylian A. Madeira
Rivelino P. Coelho

Learning theories

Connectivism in the Semi-Presential Teaching Modality

ScienciaScripts

This book is a translation from the original published under ISBN 978-620-2-04863-7.

Publisher:
Sciencia Scripts
is a trademark of
Dodo Books Indian Ocean Ltd. and OmniScriptum S.R.L publishing group

120 High Road, East Finchley, London, N2 9ED, United Kingdom
Str. Armeneasca 28/1, office 1, Chisinau MD-2012, Republic of Moldova, Europe
Printed at: see last page
ISBN: 978-620-7-23942-9

SUMMARY

PRESENTATION

Dear Readers,

We invite everyone to permeate the advances in Information and Communication Technologies (ICTs) from a perspective of educational change, especially in the form of implementing semi-presential courses in higher education. Connectivism emerges in this context as a new educational approach. This theory proposed by George Siemens and Steven Downes (2004) points out that knowledge is distributed in a network of connections and that learning therefore consists of the ability to build these networks and circulate in them, thus developing the ability to reflect, decide and share. In this investigative journey, new perspectives imposed by the connectivist theory emerged that were necessary for a better understanding: the constant traffic through learning theories for the purpose of comparison, the heutagogical and andragogical models of learning, knowledge in networks and the instructional design of distance learning courses. Thus, the approach relied on the theoretical contribution and ideas of Siemens, Downes, Zezina Belan, Vani Moreira Kensky, Andrea Filatro, Jose Manoel Moran, Augusto de Franco, Jean Piaget, Skinner and others who provided an interdisciplinary character to the research.

INTRODUCTION

Contemporary society has evolved significantly and so relationships with elements native to this evolution are also changing. The advance of Information and Communication Technologies (ICTs) has also brought about these changes in higher education, especially in the way face-to-face, distance and semi-presential courses are implemented.

Today, the role of the "e-teacher" and the "e-student" in this new environment is being discussed, as they are faced with the demands of absorbing new demands and teaching methods, but technological supports have not been discussed with due concern.

Connectivism emerges in this context as a new educational approach. This theory points out that knowledge is distributed in a network of connections and that learning therefore consists of the ability to build these networks and circulate in them, thus developing the ability to reflect, decide and share; and the student is able to conduct their learning more autonomously without the presence of the teacher.

Online learning environments, which used to be predominantly pedagogical, have been reinforced by the emerging models of open learning (Heutagogy), adult learning (Andragogy) and network learning (Connectivism).

For the development of the research, the first part addresses the learning theories - behaviorist, cognitivist and constructionist - indicated by Siemens (2004) as the most used in instructional designs. And also, the main focus of the research, the connectivist theory pointed out as being the learning theory of the "digital age".

In the second part, in the light of the authors, the aim is to study the instructional design models used today, based on the self-directed learning model (heutagogy). In this way, some reflections are made on the tools used in instructional design. It is essential to know some of the techniques used in instructional design in order to understand some of the relationships between education and the way learning is shaped in environments that use media communication technologies as their main support. For a better understanding, we seek to organize the meanings of the words that make it up: drawing and instruction.

In the third part, we try to unveil some important concepts about the network society. It analyzes how networked knowledge is built, the processes of selecting and sharing information that lead individuals to meaningful learning. It also looks at the roles of teachers and students in this environment of new information and communication technologies.

The fourth part outlines the methodology used in the research, demarcating the time frame, as well as the subjects involved, the type of sample and the tools used to collect the data. It is proposed that an exploratory study be carried out using the bibliographic theoretical framework which, according to Marconi and Lakatos (1992), is a survey of all the bibliography already published, in the form of books, magazines, single publications and the written press, and currently with material available on the internet. This research involved the use of *online* questionnaires and systematic observation.

The aim was to investigate the ways in which Connectivist Theory could be used in a virtual *online* learning environment (MOODLE). To this end, a questionnaire proposed by Bertolin and Marchi (2010) was used to evaluate semi-presential teaching through systems of indicators by categories, which were classified as Teacher, Student and Support Staff categories.

The fifth part focused on the results and analysis of the data tabulated using the information obtained from the respondents who took part in the survey.

At the end of the paper are the conclusions and considerations in which the authors' ideas and opinions are connected to the research data in order to formulate and answer the research problem, thus providing the conjugations around connectivist theory when used in a semi-presential learning environment.

Thus, this research seeks to ensure a space for reflection on the formulation of proposals to investigate how connectivist theory, aligned with the heutagogical model, can be applied in instructional designs for semi-presential courses in the second semester of 2012 at the Vale do Carangola Colleges - a unit associated with the State University of Minas Gerais - as a proposal for open learning through network connections.

The writing of this book was based on the curiosity of knowing how connectivist theory can be added to the theories used in the instructional design of semi-presential courses at Faculdades Vale do Carangola. This led to the Hypothesis that connectivist theory can be used in the production of teaching material for semipresential courses in higher education.

The aim of this research was to analyze the application of connectivist theory in the traditional instructional design of semi-presential courses in higher education.

Knowing that the theories most commonly used to promote learning in virtual environments do not advocate the use of new information and communication technologies, they nevertheless serve as a basis for doing so.

It is believed that teachers/tutors are not using all the theoretical potential in the implementation of *online teaching* materials, thus compromising the learning of individuals who use this modality. It is therefore necessary to search for information so that new and more effective methods can be found that take account of this use and transform the current form of teaching into an archetype that advocates the use of new technologies and *online* communication tools.

This topic, chosen for its novelty, needs more comprehensive research that can support or redirect the paths that structure the semi-presential teaching modality and the emerging methods stemming from information and communication technologies.

1 THEORETICAL OUTLINE OF LEARNING THEORIES IN THE CONTEXT OF TEACHING MODALITIES

1.1 Behaviorist learning theory

The aim here is not to go into detail about the aspects and theories presented, but rather to give an overview of what has been presented in the course of the readings.

From a historical perspective, the human sciences are demarcated by three conceptions of the world and man that have distinct characteristics: the cosmocentric, the theocentric and the anthropocentric.

In the first conception, the cosmocentric one, men were mere spectators to the decisions of the Cosmos, which governed everything. It is a vision of abstract man, without history and subjugated to a metaphysical morality.

In the second, the theocentric, the foundations that explain all things, including the production of knowledge, were located outside man, who, deviating from the imperatives of the Cosmos, were incorporated by the designs of a divine order, which governed everything.

Behens (2007) reports that from the 13th to the 15th century, the Renaissance emerged, a period characterized by a refusal to accept the focus on myth (Prehistory), reason (Ancient Greece) and faith (Middle Ages) as sources of knowledge. Man needed to free himself and take the process of knowledge into his own hands. At this point in history, Behrens (2007. p. 57) states that the Theocentric view tended to be overtaken by the Anthropocentric view.

In the anthropocentric view, between the 15th and 16th centuries, there were radical transformations in the European world resulting from mercantilism, which led to the discovery of new lands and the accumulation of wealth - the formation of a new economic and social organization. The new social order called for the liberation of man from his condition as a subject imprisoned by the immutable laws of the universe and defended the possibility of unraveling Nature (FREITAS, 2000).

Behrens (2007) explains that in the second phase of Modernity, from the end of the 18th century to the beginning of the 19th century, mathematical tests were used to try to understand the individual and their personality and intelligence. At this point, psychology became a science, separating itself from philosophy, and various psychological theories emerged.

The evolutionary line of psychology, therefore, is a constant oscillation between the two antagonistic theoretical strands: at one moment it points towards the principles of an objectivist tendency, which gives priority to what is external; at another, towards a subjectivist tendency based on consciousness. The aim of both strands, in turn, is to meet the requirements of the historical moment in which they originate.

In the 20th century, the behaviorist current of thought emerged as an objective and scientific methodological approach based on experimental proof, but concentrating on human beings, studying and analyzing their behavior (learning, stimulus and response reactions, etc.), concentrating on the concreteness of the facts and not on subjective and theoretical concepts such as sensation, perception, emotion, attention (CHIAVENATO, 1997).

From the English *behaviourism,* the term *behaviour* means conduct, behavior. Behaviorists work on the principle that the conduct of individuals is observable, measurable, giving rise to the theory of behavior or behaviorism developed initially by the American psychologist Burrhus Frederic Skinner, considered one of the fathers of behavioral psychology.

According to Block et al. (1999), the term behaviorism was introduced by the American John B. Watson in a 1913 article entitled "Psychology: As Behaviorists See It". Although John B. Watson considers himself to be the founder of Behaviorism, the name most remembered when talking about this theory is that of Skinner.

Behaviorism studies psychological events based on behavioral evidence and presents itself as an objective psychology as opposed to subjectivism; it is a theory that understands psychology as a science of behavior and not of the mind Graham (2007). In this view, behavior is explained without reference to mental events, since these can be translated into behavioral concepts.

However, for a better understanding of the set of ideas on behavior analysis, the following are the characteristics related to the three main models of Behaviorism since, since its first generation, the notion of "behavior" has undergone important transformations.

1.1.1 The first generation: methodological behaviorism

John B. Watson, with the publication of "Psychology: As the Behaviorists See It", inaugurated, in 1913, the term that would come to denote one of the most expressive theoretical trends still in force: Behaviorism. By presenting behavior as the object of study for psychology, Watson established an object of study that was "[...] observable and measurable, whose experiments could reproduce different conditions and subjects" (p.

45).

Methodological behaviorism is based on realism. Realism defends the idea that there is a real world, which takes place in the real world, and it is from this external - objective - real world that the internal - subjective - world is constituted. Paradoxically, we only rely on the internal experience provided by our senses. This is because the external, objective world is not directly accessible. Therefore, the senses only provide sensory data about that real behavior which is never known directly (BAUM, 1999).

1.1.2 The second generation: radical behaviorism

In 1945, B.F. Skinner introduced Radical Behaviorism, in which he defended the experimental analysis of behavior. In contrast to Watson's methodological, realist Behaviorism, Skinner's thinking adopts the principles of pragmatism, as he is concerned with the functionality of the observable, measurable real object, and not with the existence of a real object behind these effects (BAUM; FURTADO, 1999).

The behaviorists of this generation try to break away from the objective-subjective world duality; instead of relying on methods, they adopt concepts and terms.

[...] the terms we use to talk about behavior not only allow us to understand it, but also define it: behavior includes all the events we can talk about with our invented terms (SKINNER, 1974 *apud* BAUM, 1999, p. 45).

The radical behaviorists postulated by Skinner therefore understand all natural events, subject to access, including public and private events; and exclude fictitious ones - which cannot be accessed. The mind and processes, as mental causes of behavior, are considered fictitious and therefore constitute "terms" that should be avoided. Radical behaviorists thus assume that the causes of behavior lie in heredity and in the past and present environment.

Skinner (1974) emphasizes:

Radical behaviorism restores a certain kind of balance. It doesn't insist on truth by consensus and can therefore consider private events inside the skin. It doesn't consider such events unobservable and doesn't dismiss them as subjective. It simply questions the nature of the observed object and the reliability of the observations (p. 19).

Radical behaviorism sets out to explain animal behavior through the model of selection by consequences. For Skinner, most human behavior is operant conditioned. The term radical is explained by the analysis focused on the root of a specific behavioral phenomenon.

1.1.3 The third generation: social behaviorism

Staats (1980), corroborating the idea that revolutions are generally made against those in power, disagrees, however, with extremist observations that overthrow the old order and don't bother to separate what is right and wrong in it, [a]takes up his position as a representative of the 3rd generation of behaviorists, whose aim was, in addition to transposing the two previous generations of behaviorists, to move towards an explanation of behavior that takes into account human-environment interaction in a broader way.

Roughly speaking, when we look at the characteristics of the three main models of Behaviorism: methodological behaviorism, which is based on realism and whose greatest expression is Watson; radical behariorism, which is based on the principles of pragmatism and whose greatest representative is Skinner; and social behaviorism, which was born with Staats, in opposition to the two previous programs because they considered them to be closed systems and therefore reductionist, the aim is to demonstrate, albeit briefly, that behaviorism is directed towards a more humanistic conception of behavior.

It is in Brazil, however, that Skinner's Behaviorism influences psychologists, as it does in many countries around the world where American psychology is more present Bock (2008). And yet, because the English term *behavior* means "behavior", in Brazil Behaviorism is used as well as Behaviorism, Experimental Behavior Analysis, among others.

The pillar of Skinnerian thinking lies in the formulation of operant behavior, which has respondent behavior as its historical basis. Bock (2008) makes some statements and gives some examples of this behavior.

Reflex or respondent behavior is what we usually call "non-voluntary" and includes responses that are elicited (or produced) by antecedent stimuli from the environment. As an example, we can cite the contraction of the pupils when a strong light shines on the eyes, the salivation provoked by drops of lemon placed on the tip of the tongue, the shivering of the skin when cold air hits it, etc (BOCK, 2008, p. 59).

According to Bock (2008, p. 63-67), the mechanisms of operant conditioning that Skinner considers important are:

• Positive reinforcement or reward: any event that increases the future probability of the response that produces it;

• Negative reinforcement: any event that increasesthe future probability of the response that removes or attenuates it;

- Extinction or absence of reinforcement: responses that are not reinforced are unlikely to be repeated (ignoringstudents' misconduct, the expected conduct should be extinguished); and

- Punishment: responses that are punished can lead to undesirable consequences (a late punishment for a student could have no effect at all).

Bock (2008) also points out that operant conditioning covers a wide range of human activities and behaviors from the cradle to the most sophisticated behaviors exhibited by adults.

In this respect, according to Bock (2008 *apud* KELLER, 1973) operant behavior includes all the movements of an organism which can be said to have an effect on or do something to the world around it at some point. Operant behavior acts on the world, so to speak, either directly or indirectly.

For Moreira (2009), in the behaviorist theory created by Skinner, teaching is achieved when what is taught can be placed in conditions of control under observable behavior; thus, it is achieved when unwanted behavior is punished and the desired behavior is reinforced and encouraged with stimuli until it becomes automatic.

The main objections to Skinner's system are that it focuses exclusively on the effect caused by a given learning task and ignores the internal cognitive process that has taken place in the learner.

1.2 Cognitivist learning theory

Cognitive Psychology can be defined by Sternberg (2000, p. 38) as "[...] the study of how people perceive, learn, remember and ponder information". On the other hand, Cognitivism can be described as a psychological perspective suggesting that the study of how people think will lead to broad *insight* into much of human behavior (STERNBERG, 2000).

Around the 1970s, a considerable number of psychologists began to reject the classic model of stimulus and response - Behaviorism - which tended to ignore human activities such as reasoning, planning, decision-making and communication, and increasingly sought to understand what went on inside the human mind. This gave rise to the movement known as cognitivism. Judith Grenne (1976 *apud* PENNA, 1984, p. 3) conceptualizes this movement as a "[...] current that deems it impossible to understand the *input/output* relationships registered in human behavior without taking into account the strategies and rules that a given subject is using when faced with an impasse".

Understanding how people think has stimulated research in this field for a long time. And all the time, theories, models and proposals for cognitive structures emerge to try to explain something that to the less knowledgeable may seem very difficult to explain.

Moreira and Masini (1982) elaborate on the concept of cognition:

It is the process through which the world of meanings originates. As people situate themselves in the world, they establish relationships of meaning, in other words, they attribute meanings to the reality in which they find themselves. These meanings are not static entities, but starting points for the attribution of other meanings. The cognitive structure (the first meanings) then originates, constituting the basic anchor points from which other meanings derive (p. 3).

Cognitivism counters and emphasizes what is ignored by behaviorist theory, which focuses on biological aspects and human behavior through the analysis of the mind. It is an approach that involves scientifically studying learning not as a mechanically internalized factor of the individual, but as a product of the environment, people and external factors, thus creating a network of meanings.

According to Bock (2008), the cognitivist theory approach differentiates mechanical learning from meaningful learning, defining mechanical learning as the type of learning of new information, with little or no association with concepts already existing in the cognitive structure or when the material learned does not find an echo in the subject's biology, because it does not interact with the relevant concepts existing in the cognitive structure, being stored in an arbitrary and literal way. Thus, the knowledge acquired is arbitrarily distributed in the cognitive structure, without being linked to specific concepts.

The author also defines meaningful learning, stating that this type of learning takes place when new content (ideas or information) is related to relevant, clear concepts that are available in the cognitive structure and are thus assimilated by it. He calls these available concepts anchor points or subsumers for learning (BOCK, 2008, p. 135).

The concept of subsumption defended by Ausubel (2003 *apud* MOREIRA, 1993) is a reference or proposition that already exists in the learner's cognitive structure which serves as an interface for new information and allows the individual to attribute meaning to it. In this respect, meaningful learning is characterized by interaction and not by a simple association of information.

Moreira (2006) also agrees:

[...] meaningful learning is a process by which new information is related in a substantive, non-arbitrary and non-literal way to a relevant aspect of the individual's meaningful structure. The new information interacts with a present cognitive structure (p. 15).

For Ausubel (1963), meaningful learning is the human mechanism par excellence for acquiring and storing the vast amount of ideas and information represented in any field of knowledge. Ausubel (1963) says that learning has three essential advantages over memorizing: firstly, knowledge that is acquired in a meaningful way is retained and remembered for longer; secondly, it increases the ability to learn other content more easily, even if the original information is forgotten; thirdly, it is necessary to modify the subject's cognitive structures as a result of learning meaningfully.

With regard to expanding and changing the learner's structures, Piaget (1997) postulates that it is necessary to provoke disagreements or cognitive conflicts that represent imbalances from which, through activities, the student can rebalance themselves, overcoming the disagreement by reconstructing knowledge. To achieve this, learning must not be excessively simple, which would lead to frustration or rejection.

In turn, mechanical learning occurs when the learner memorizes formulas, laws and schemes that they will soon forget. It is also characterized by the inability to use and transfer this knowledge. Ausubel does not distinguish between them (meaningful and mechanical), thinking of them more as a set of continuous situations (AUSUBEL, 1978; MOREIRA, 2006).

According to Ausubel's theory, learning can take place by discovery or by perception. In discovery learning, the student learns alone and has to discover the principles or by relationship and this can happen when solving problems. In perceptual learning, the student receives finalized information and their job is to act on this material in order to relate it to ideas in their cognitive structure.

According to Filatro (2003), the first signs of the so-called cognitive revolution in the fields of instructional design were launched by Piaget in the 1920s, but became evident in the 1950s-1960s. The author compares cognitive structures with the human mind, stating:

The intense emphasis on mental schemas that characterizes cognitivism coincides historically with the development of computer technology from the mid-twentieth century onwards, revealing an understanding of mental processes that resembles the operating patterns of computing machines. The comparison of the human mind with the basic structure of a computer established the information processing model as a new paradigm (p. 81).

Filatro (2003) also states that

According to this model, the mind, like a computer, initially receives sensory records that are processed and stored in the form of schemas, which are activated and restructured in the learning process, and retrieved when necessary (p. 81).

In this process of building knowledge, from a multimedia perspective, Moran (1998) infers that this process is freer, less rigid, with more open connections that go through the sensory, the emotional and the rational organization. In this way, the organization of thought and knowledge becomes provisional and easily mutable, creating instant convergences and divergences, with multiple processing and immediate response.

Moran (2000) points out that the individual learns through interest, through need, when he experiences, feels, establishes bonds, discovers new possibilities and integrates the sensory, the rational, the emotional, the ethical, the personal and the social. Finally, he learns when he interacts with others and the world and then when he internalizes, when he turns inwards, making his own synthesis, the re-encounter of the outside world and personal re-elaboration.

Cognition, therefore, as a vision of how the individual internalizes and externalizes the information they receive, transforming it into meaningful knowledge, can involve a large behavioral structure when transferred to the field of instructional design in a virtual learning environment.

1.3 Constructivist learning theory

For Boyle (1997), Constructivism has been the theoretical approach most often used to guide the development of computerized teaching materials, especially multimedia learning environments. But this does not mean that these approaches have to permeate as the only trend in media teaching.

This statement is also confirmed by Siemens (2004) who adds the theoretical approaches of Behaviorism and Cognitivism, and makes it clear that all approaches are welcome in order to achieve the goals of producing knowledge, whether in an online or face-to-face environment.

According to Biaggio (1976), the biologist, doctor of philosophy and epistemologist Jean Piaget was the scholar who stood out most in studies on Constructivism. Considered one of the great names in developmental psychology, he developed a theory called Genetic Epistemology or Psychogenetics to explain how human intelligence develops.

According to Piaget's Genetic Epistemology or Psychogenetics, the individual matures his or her intelligence in constant interaction with the environment, despite being affected by various biological factors.

For Piaget (1990),

[...] knowledge does not originate either from a self-conscious subject or from the constituted objects (from the subject's point of view) that would be imposed on it: it results from interactions that take place halfway between subject and object, and which therefore depend on both at the same time, but by virtue of a complete undifferentiation and in exchanges between distinct forms, (p. 8).

According to Altoé (2005), subjects are seen as constructors of their knowledge, since, through interaction with the environment and based on previous experiences, they formulate propositions in an attempt to resolve new situations. During the procedure, cognitive constructions emerge in consecutive movement and which, driven by the search for balance, are capable of producing new mental structures.

The constructivism defended by Piaget is based on the principle that knowledge is not something that is finished, but rather a process in constant construction and conception as a result of interaction with the environment, from which the individual's personality is formed. Thus, knowledge is a building erected through action, elaboration and the generation of learning that is the product of the individual's connection with the material and social context in which he or she lives, with the symbols produced by the individual and the universe of interactions experienced in society.

The role of the teacher in the Piagetian line of thought is to observe the student, find out what their previous knowledge is, what their interests are and, based on this vision, try to provide elements for the student to build their knowledge. The teacher creates situations for the student to arrive at knowledge.

Another scholar who also underpinned constructivist theory was Lev Vygotsky. According to Carretero (1997), one of his essential contributions was to conceive of the subject as an eminently social being, in line with Marxist thinking, and of knowledge itself as a social product. For the authors who followed the constructivist line, this meant that learning was not considered to be an individual activity, but rather a social activity.

Bock (2008) explains that,

When studying learning, Vygotsky emphasizes and highlights the importance of social relationships in the process. All suggestions for teaching methods and procedures must value and include relationships with peers. In learning, contact with others, with the already humanized and cultural world, is an essential factor. Development and achievement are the result of these interactions (p.144).

In this way, it is understood that both Piaget and Vygotsky take into account the interference of biological, cognitive, emotional and social aspects when dimensioning the learning process and defend interaction as an essential factor in the process of constructing knowledge. This postulate leads to the concept that interactions are always present in learning.

1.4 A new approach: Connectivist Theory

The world is governed by a new society, the connected one. The term connection will be understood here, as Filatro (2003) informs us, as a moment when many come together around the same idea.

Connectivity is when two or more people come together mentally, interact, talk or collaborate. With the help of telegraphs, radios, telephones or digital communication networks, these people can be in different, distant places. The advance and expansion of the use of the Word Wide Web (WWW) has transformed the possibilities of connectivity between people (p.102).

In this new connected world, where technologies allow for an unlimited flow of information, there is a need to keep up with this evolution because, as Siemens (2004) points out, the world used to live in a time when knowledge was measured in decades and now it is measured in months and years. What you learn today may no longer be true tomorrow. There is also the impact of informal learning through society and social networks, as a continuous process in which -knowing where" is fundamental, i.e. knowing how to find the knowledge you need at a given moment, to the detriment of -knowing how" and -knowing what".

It is understood that due to the amount of information available in the networked society, it is essential to learn and systematize ways of acquiring information, selecting it and choosing it as a valid, safe and reliable source. This new vision leads to the observation that learning is based on the principle that knowledge is distributed and therefore not transferable.

In 2004, professor and director of the Learning Technology Center at the University of Manitoba (Canada) George Siemens, together with Steven Downes, a participant in a study group at the Information Technology Institute for distance learning in Canada, both explorers of the pedagogical possibilities of new information and communication technologies, proposed through scientific articles, book chapters and *online* media, a new theory of learning, called Connectivism, presented as a new teaching-learning paradigm.

Siemens (2004) points out that Behaviorism, Cognitivism and Constructivism are the most widely used theories in the creation of instructional learning environments. However, these theories were developed at a time when there was no impact of technologies. These technologies have favored new forms of social and institutional communication and conceptualize learning as occurring within the person, not considering that which occurs through living in society.

According to Siemens (2004), technology has reorganized the way we live, the way we communicate and the way we learn; in this context, learning can take place in a variety of ways, with emphasis on informal learning through communities of practice, personal networks and also work-related activities.

Connectivism is a recent theory and characteristic of the information age, which is why it requires a different look at its possibilities for application in education.

SIEMENS (2004) states that:

Connectivism presents a model of learning that recognizes the tectonic shifts in society, where learning is no longer an internal, individual activity. The way a person works and functions is altered when new tools are used. The field of education has been slow to recognize both the impact of new learning tools and the environmental changes in which learning has taken place. Connectivism provides insight into the skills and learning tasks needed for learners to flourish in the digital age (p. 8).

According to (Siemens, 2008, p. 8) knowledge is distributed through an information network and can be stored in a variety of digital formats: -Learning and knowledge rest on the diversity of opinions".

According to the aforementioned author, "[...] connectivism is the integration of principles explored by chaos, networks and theories of complexity and self-organization" (p. 8). Siemens (2004) also describes some of the theory's characteristics:

- learning and knowledge are based on a diversity of opinions;

- learning is a process of connecting specialized nodes or sources of information;

- learning can reside in non-human devices;

- the ability to know more is more critical than what is currently known;

- it is necessary to cultivate and maintain connections to facilitate continuous learning;

- the ability to see connections between areas, ideas and concepts is a fundamental skill; and

- currency, accurate and up-to-date knowledge) is the intention of all connectivist learning activities (p. 6).

This notion of a new learning theory based on systematized networks and complex, changing environments has not been without its critics. Bill Kerr (2007) has postulated that Connectivism is an unnecessary theory, because according to him, existing theories cater well for current learning processes based on new technological models. Like Plon, Verhagen (2006), in his article (*Connectivism: a new learning theory?*), provides some specific arguments for the ineffectiveness of a theory based on "unfounded philosophies". Furthermore, Verhagen (2006) is not convinced that learning can reside in non-human devices.

Verhagen's (2006) criticism focuses on three areas:

1. Is Connectivism a learning theory or a pedagogy?

2. Are the principles advocated by Connectivism present in other learning theories?

3. Can learning reside in non-human mechanisms?

Verhagen (2006) does not classify Connectivism as a theory, going so far as to say that it would be better classified as a pedagogical and curricular perspective, since theories deal with questions pertaining to the level of instruction, "how individuals learn" and Connectivism, in turn, in his view, reaches the curricular level, what is learned and why is it learned (KOP; HILL, 2008).

Kop and Hill (2008) have the same critical view as Kerr and Verhagen, that the principles of Connectivism do not justify it as a learning theory; however, they recognize that the theory in question contributes to the current context of paradigm shifts, in which the student has increasingly acquired a position of autonomy in the learning process.

Siemens (2006), in response to Plon Verhagen's criticism, comes up with a very well-founded article - *Connectivism: Learning Theory or Pastime of the Self-Amused?* - in which he reaffirms the postulates of Connectivism, justifying it with an analysis of learning theories. In this article, Siemens admits that there have been developments due to technology, compared to his original article, and points out 5 fundamental issues to distinguish a learning theory.

1. How does learning take place?

2. What factors influence learning?

3. What is the role of memory?

4. How does the transfer take place?

5.   What types of learning are best explained by this theory?

After an analysis of the perspectives on what behaviorist, cognitivist and constructivist theories postulate about knowledge and learning,

Siemens tries to explain some of the aspects relating to Connectivism and, based on a summary table (Table 1), has produced the differences and also similarities between the various theories, as well as answering whether or not Connectivism should be considered an autonomous theory.

Properties	Behaviorism	Cognitivism	Constructivism	Connectivism
How does learning take place?	Black box - focus on observable behavior	Structured, computational	Social, meaning constructed by each learner (personal).	Distributed in a network, social, technologically enhanced, recognizing and interpreting patterns.
Influencing factors	Nature of reward, punishment, stimuli.	Existing *schemas*, previous experiences.	*Engagement*, participation, social, cultural.	Network diversity.
What is the role of memory?	Memory is the inculcation (*hardwiring*) of repeated experiences where reward and punishment are more influential.	Encoding, storage, -*retrieval*.	Previous knowledge remixed for the current context.	Adaptive patterns, representative of the current state of the networks.
How does the transfer take place?	Stimulus, response.	Duplication of knower knowledge constructs.	Socialization.	Connection (addition) with *nodes*.
Types of learning better explained	Task-based learning.	Reasoning, clear objectives, problem solving.	Social, vacancy	Complex learning, rapidly changing core, diverse sources of knowledge.

Chart 1 - Learning theories Source: Siemens, 2006, p. 36.

This comparative analysis makes it possible not only to justify Connectivism as a learning theory by answering the five fundamental questions, but also to highlight the limitations of existing theories for the knowledge age, characterized by information and communication technologies. For Siemens (2006).

As an answer to the question of whether learning can reside in non-human devices, Siemens (2004, p. 5) explains that

Learning is a process that takes place within nebulous environments where the core elements are changing - not entirely under people's control. Learning (defined as actionable knowledge) can reside outside of ourselves (within an organization or database), is focused on connecting specialized sets of information, and the connections that enable us to learn more are more important than our current state of knowledge.

From this perspective, it can be seen that the power residing in non-human devices is a variant of the effect that is evident in the concept of learning. Siemens (2004) believes that knowledge is internalized in the individual and all that is needed is for a trigger to be activated for it to be converted into learning, which he calls "actionable knowledge". It can then be understood that learning can reside in non-human devices, such as an organization, a cell phone or a database.

Siemens (2008, p. 8) also agrees that Connectivism offers some central points that give it its originality.

1.   Connectivism is the application of network principles to define both knowledge and the learning process. Knowledge is defined as a particular pattern of relationships and learning as the creation of new connections and patterns, on the one hand, and the ability to maneuver through existing networks and patterns, on the other.

2.   Connectivism deals with the principles of learning at various levels - biological/neural, conceptual and social/external.

3.   Connectivism focuses on the inclusion of technology as part of our distribution of cognition and knowledge. Our knowledge resides in the connections we create, whether with other people or with sources of information such as databases.

4.   While other theories pay partial attention to context, connectivism recognizes the fluid nature of knowledge and connections based on context.

5.   Understanding, coherence, sensemaking, meaning: these elements are prominent in constructivism, less so in cognitivism, and are absent in behaviorism. But connectivism argues that the rapid flow and abundance of information elevates these elements to a critical level of importance.

Siemens (2004, p. 8) also argues that learning theories are concerned with the actual process of learning, not the value of what is being learned.

In a networked world, the exact kind of information we acquire is by exploring its importance. The need to assess the importance of learning something is a meta-skill that is applied before the learning itself begins. When knowledge is subject to parsimony, the process of evaluating importance is assumed to be intrinsic to learning. When knowledge is abundant, rapid evaluation of knowledge is important. Additional concerns arise from the rapid increase in information. In today's environments, action is often necessary without personal learning - that is, we need to act by seeking information outside of our primary knowledge. The ability to synthesize and recognize connections and patterns is a valuable skill.

In this sense, it can be said that the information that the individual receives in an information network needs to be processed, because the rapid flow and its abundance elevate the learner's need for it to critical importance. Connectivism finds its roots in the diverse sources of information, rapid changes and perspectives, in which it is necessary to find a way of filtering and making sense of the chaos.

Landauer and Dumais (1997 *apud* SIEMENS, 2004) explore the phenomenon that people have much more knowledge than appears to be present in the information to which they have been exposed. They state that the simple notion that some knowledge domains contain a vast number of weak interrelationships which, if properly exploited, can greatly amplify learning through a process of inference.

Downes (2005) contributes by describing knowledge as a network phenomenon: to know something is to be organized, to display patterns of connectivity and to learn is, in this context, to acquire these patterns, and this applies to both a person and a society. In an article published the following year, Downes (2006) explains that learning takes place in communities and that the practice of learning is participation in the community itself. He adds that a learning activity is the essence of a conversation between the learner and other members of the community. This phenomenon, seen through the lens of Web 2.0, means that communication is not only contained in words, but also in images, videos, in short, in the media.

In this way, the Characteristics of Connectivism described by Siemens (2005) and Downes (2006) as well as by Verhagen (2006) and Kerr, Kop and Hill (2008) are aimed at individuals with an aptitude for autonomous learning, suggesting that they take responsibility for managing their learning and use media tools, giving rise to further studies.

1.5 Learning theories and their pedagogical aspects in *blended* learning modalities

The modern educational scenario is characterized by a strong convergence: the resumption and development of distance education from new perspectives of its use with the addition of new communication technologies to break down paradigms and prejudices.

For Borges (2005), the emergence of this type of teaching can be traced back to the mid-19th century in Russia and Belgium, with the first correspondence courses. Then, in the 1970s, we had the generation of distance education based on teleeducation, which used mass media such as radio, television and printed material. In the 1980s, audio and video were incorporated. In the following decade, satellite networks, the computer, electronic mail, the use of the Internet and computer programs designed especially for education radically changed this modality (LITWIN, 2001).

According to Quevedo (2011), the approval of the National Education Guidelines and Bases Law in 1996 recognized distance learning as a valid educational modality for all levels of education in Brazil. From then on, distance learning postgraduate courses began to be offered throughout the country. In 1998, the Ministry of Education authorized distance learning for undergraduate courses.

In April 2001, the Ministry of Education published decree 2, which authorized all universities, colleges and technology centers to offer distance learning for up to 20% of the curricular load of recognized and ongoing courses. This gave rise to the semi-presential mode of teaching.

According to Voigt (2002), we can't talk about semi-presential teaching without establishing a direct connection with distance and face-to-face teaching, as the latter promotes an understanding of the other two types of teaching. We know that in what we know as face-to-face teaching there is direct contact between student and teacher, the latter being present in the classroom and being the holder of knowledge.

Voigt (2002) also states that

Although it has only attracted attention in recent years, distance education is nothing new in Brazil and worldwide. In Brazil, the first documented experience dates back to 1904. Early distance learning was based on handouts sent by post or published in newspapers. Later, other communication technologies were used: radio, TV and, more recently, the internet. The emergence of modern communication and information technologies (ICT) has led to a substantial growth in distance learning in recent decades (p. 46).

Voigt (2002) also points out some important elements that distinguish distance learning from face-to-face teaching, with the use of technological means: information and communication technologies, the use of technologies, the physical separation between teacher-student and student-students; the redefinition of the concept of presence; autonomous learning and meticulous planning of the teaching-learning process.

It can be seen that the change from the face-to-face model to the distance model requires significant transformations both in the representations of students and teachers and in the institutional structure. These transformations, seen by some professionals as a paradigm, have led to the emergence of a third modality: hybrid or *blended learning* (TORI, 2009), semi-presential and bimodal systems (MORAN, 2004).

Torin (2009) begins his discussion by arguing about the convergence of face-to-face and virtual *learning,* known as *blended learning,* with the following comment:

Two learning environments that have historically developed separately, the traditional face-to-face classroom and the modern virtual learning environment, are discovering that they are mutually complementary. The result of this encounter is hybrid courses that seek to take advantage of each modality, considering context, cost, pedagogical suitability, educational objectives and student profiles (p. 121).

Moran (2004) seems to agree with Torin's (2009) idea when he describes semi-presential teaching as two distinct teaching modalities - a modality that mixes face-to-face activities with distance activities, facilitated by the use of information and communication technologies, and calls it a bimodal system (MORAN, 2004).

Semi-presential teaching can therefore be seen as a way of democratizing knowledge, with the application of new communication and information technologies, with a view to reshaping the teaching-learning process, as well as the role of teachers and students, without dispensing with the importance of constant face-to-face meetings.

Distance education was regulated in Brazil by Law 9394 of December 20, 1996 and by Decrees 2.494 of February 10, 1998 and 2.561 of April 27, 1998. However, the semi-presential teaching modality was only regulated by the Ministry of State for Education (BRASIL, 2005) on December 10, 2004, through decree 4.059.

According to this decree, higher education institutions (HEIs) can introduce subjects that use the semi-presential modality into the curriculum of their recognized courses.

Ordinance 4.059/04 Art. 1 paragraph 1 of the MEC describes semi-presential education as

[...] any didactic activities, modules or teaching-learning units centered on self-learning, and with the mediation of didactic resources organized in different information supports that use remote communication technologies (BRASIL, 2005, p.1).

Moran (2004) deals with the issue in a simpler but no less important way, describing that semi-presential education takes place partly in the classroom and partly at a distance, through the use of technology. For the author, distance education may or may not have face-to-face moments, however, it fundamentally takes place with teachers and students physically separated in space and/or time, and may be brought together by means of communication technologies.

In order for semi-presential education to be practiced at HEIs, the interested courses must be duly recognized by the MEC, and the distance learning workload may not exceed 20% (twenty percent) of the total course workload.

This modality, like face-to-face or totally distance learning, requires prior planning, so that the subject can be directed in such a way as to lead the student to self-learning, and it can also provide support to help in the process of breaking away from the traditional concept of education, which is based on the transmission of content, centered on the figure of the teacher, whose predominant type of communication is one-way, one-to-all communication. Distance education, on the other hand, uses digital technologies to build a one-to-many communication model.

It is worth emphasizing that this is semi-presential *online* education, which is based on collaborative learning, on the process of co-authorship, in which the teacher mediates and guides the learning activities (SILVA, 2001, 2003; BORGES; FONTANA, 2003).

It is not possible to choose a theory that serves as a model for this or that modality, since its particularities do not allow for a centralized choice, and since the changing and evolving nature of paradigms leads to the belief that more research is needed in this area, the main objective of which is to promote the thinking and rethinking of semi-presential education, which will contribute to the construction of the "design" of this hybrid teaching modality.

In this way, all the theoretical currents used in *online* teaching are intrinsically linked to a new educational model, based on the experiences and autonomy of the target audience. As the currents used to date did not have the use of new technologies as their main reference, new paradigms are being established at a time when there is a migration from the classic style to new emerging styles, aligned with new concepts and technologies offered by the internet.

2 THE INSTRUCTIONAL DESIGNER

2.1 The instructional designer in online education

In this chapter we will reflect on the meaning of the term "instructional *designer*" and its correlates. To this end, it is essential to know the role of this educational actor in order to understand some of the relationships between education and the way learning is shaped in environments that use media communication technologies as their main support. For a better understanding, we tried to organize the meanings of the words that make it up: design and instruction.

2.1.1 Instructional designer in the various teaching modalities

In recent times, we have seen the growing evolution of technological media - TV, radio, newspapers, cell phones, DVDs - today, they are all converging on the computer. Access to the world is just a *mouse* click away. Electronic messages are sent and received in a matter of seconds from anywhere in the world to anywhere else. People are increasingly connected to each other through digital social networks whose sources of information seem endless.

In this context, the computer has become an extremely important tool. With this resource, the school breaks down the boundaries of time and space, thus expanding the universe of research in all areas of human knowledge.

However, this fascinating technology is not solely responsible for the process of change that the world and educational institutions are going through. In order to take advantage of the benefits that computer resources can offer, we need, first and foremost, training and a change in behavior, as well as new curricular projects.

Ongoing teacher training is necessary in several areas, starting with their own mastery of the content, which is often unsatisfactory. In the pedagogical field, for example, teachers have little opportunity to learn about new approaches that can support appropriate methodologies that facilitate learning. Another area where difficulties are faced is the introduction of technology into teaching practice, even when the school has technological resources. Kenski (2003) warns that

This professional needs to have time and opportunities to become familiar with the new educational technologies, their possibilities and their limits, so that in practice they can make conscious choices about the use of the most appropriate ways of teaching a given type of knowledge, at a given level of complexity (p. 48).

All the teaching methods now facilitated by information and communication technologies, especially the Internet, are becoming even more attractive as the need to build environments and tools that enable learning via the Internet brings the opportunity to rethink the educational paradigms that have been used in conventional formal education, a fundamental issue in any pedagogical project of technological innovation.

In this sense, *online* education presents itself as a very viable alternative to meet the constant need for continuing education for in-service teachers. The facilities offered by information and communication technologies (ICT) can serve as fundamental resources for education systems, enabling interaction between training centers, students and professionals, regardless of time and space.

From a broader and less instrumental perspective, ICTs can represent new resources to support learning; the use of these technologies also strengthens a recent movement within the theory and practice of instructional design, which proposes the adoption of a new way of planning teaching and learning.

The term "instructional *design*" has been used in various senses and meanings and refers to the planning of teaching. In its literal translation, *design* means planning and instructional refers to the idea of teaching.

According to the Brazilian Distance Education Association (ABED), Instructional *Design* (ID) is an area of Educational Technology defined as "[...] the study and practice of facilitating learning and improving performance through the creation, use and management of appropriate technological processes and resources" (AECT, 2008, p. 3).

Romiszowsky (2005) looks back at this discipline 50 years ago, at the height of the use of programmed instruction. In this period, considered the "era of behaviorism", the concern in terms of education was with teaching technology, the application of scientific principles to teaching situations.

According to Filatro (2003, p. 55), the term "instructional designer" was originally coined in 1588, the English word meaning "intention, purpose, arrangement of elements in a given artistic pattern", coming from the Latin *designare,* "to mark, to indicate", and via the French *designer,* "to design, to draw".

The *designer's* field of research is based on understanding how information and communication technologies contribute to improving the teaching-learning process and represents an opportunity to rediscover the unique, irreplaceable, highly creative nature of education in the process of human and social development and to propose solutions. It can be understood as the planning, development and systematic use of teaching methods, techniques and activities for educational projects supported by technologies (FILATRO, 2004, p.32).

Also according to Filatro (2008), instructional *design* is defined as the intentional and systematic teaching action that involves the planning, development and application of methods, techniques, activities, materials, events and educational products in specific didactic situations in order to promote human learning based on known learning and instructional principles.

Siemens (2005) also contributes by saying that effective instructional *design* must recognize different learning domains, adapting to the background of students and subjects. Thus, Instructional *Design* is the discipline that is interested in the process of instruction, increasing the prospects for learning.

It can be seen from the authors' statements that the instructional *designer* is a professional capable of proposing solutions in an organized and systematic way (Figure 1) for learning in any educational environment, based on an analysis of the students and their needs. These solutions can take the form of face-to-face, semi-presential or totally distance learning systems.

They can also use them independently or combine the use of teaching materials in different media.

Figure 1 - Conventional instructional *design* development model

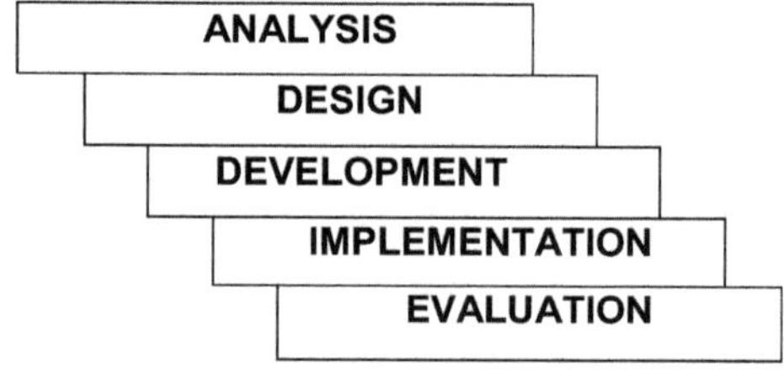

Source: Filatro, 2003, p. 70.

Filatro (2003) proposes solutions for the development of instructional design, taking into account the various educational contexts, when he presents some guiding questions, clarifying the methodology of application to didactic situations.

DI plays an essential role in bringing together fields of study in an interdisciplinary perspective - for the betterment of students, instructors and organizations, facilitating learning, reducing drop-out rates, student resistance and performance.

Thus, through a systematically structured planning process for learning objects, a concept of quality can be imprinted on the process of producing teaching material for *online* education.

The development of teaching material for *online* teaching in virtual environments requires care to ensure that the student can take full action in relation to the material, as well as the languages involved. This implies, on the one hand, that the pedagogical parameters and the so-called complementary theories, andragogical and heutagogical, are clearly defined; and on the other, that the professionals involved in producing the material reinvigorate their training based on an interdisciplinary articulation of education and communication.

By analyzing Filatro's (2003) principles and guidelines for the development of Instructional Design in the current context in which information is changing, it is clear that ID is a very important element in the construction of distance learning courses, especially considering the absence of face-to-face contact between teacher and learners.

Based on this premise, it can also be seen that DI has its precepts strongly linked to connectivist theory, since distance learning and semi-presential courses make use of new technologies and require students to have a high degree of connection; whether this is environmental, such as network connections, or personal, in which students must remain connected to specialized information sources or data storage devices.

Contemporary technological developments are strongly linked to education. According to Kenski (2003, p. 1), "[...] it is impossible to educate without technological mediation". The researcher states that the new information and communication technologies are already established in almost all contemporary social and cultural spaces. However, this evolution does not in itself produce the skills necessary for the elaboration and systematization of new knowledge.

According to Moran (2000), society has experienced changes in the way it organizes itself, produces goods, sells them, entertains itself, teaches and learns. The author also states that today's ways of teaching are no longer justified, as too much time is wasted and too little is learned, which, consequently, continues.

Lévy (1999) then summarizes the new role of the teacher in the current context as being the animator of the collective intelligence of the groups they are in charge of. He also states that teachers' activities will be centered on monitoring and managing learning, by encouraging the exchange of knowledge, relational mediation and monitoring learning paths.

Souza (2008) points to some of the tools used to facilitate cooperation, learning and the formation of user communities, which are publicly available on the Internet: "Interaction via e-mail, Discussion Lists, Forums and *Newsgroups, Online* Conversation Environments, or *Chats, Web-based* Learning Environments, Web Portals, *Web Rings* and File Sharing Servers".

In this respect, the production of didactic material requires a pedagogical rethink, including the creation of didactic-pedagogical strategies for effective learning in a new configuration, where the media must be used to support a process planned in person and executed virtually based on available methodological theoretical pillars.

In practice, it can be seen that the instructional *designer is* moving in the direction of student flexibility and autonomy, keeping up with technological and institutional changes. For this to happen, the whole team, in a multidisciplinary approach, must align itself with learning theories or, as Demo (2009) points out, live elegantly with them, making them possible options, not fixed ideas. This goes back to the idea that not just one theory should be used as a basis for producing *online* teaching material, but that adaptations should be made according to the needs of the audience and the environment.

Siemens (2002, p. 3) points out that

Compared to a human instructor, technology is less adaptable. Once an execution plan is in place, it is less likely to change according to the students' reactions. This is why the instructional designer plays an important role in bridging pedagogy and technology. Content, subject matter has to be well organized and strategies for teaching through a chosen medium have to be thought out. The instructional designer can help teachers make the best use of technology, thus ensuring successful integration.

As for the structure of semi-presential, distance learning and face-to-face courses, Carneiro et al. (2010) list some ways of organizing online courses in advance. Firstly, they suggest the use of activity maps so that the division of content and the planning of activities can be thought out in greater detail (Chart 2).

Class	Unit (Main theme)	Sub-units (Sub-themes)	Specific objectives	Theoretical activities and distance learning resources/tools	Practical activities and EaD resources/tools
A1-	What is	-	Distance Differentiating	ATV1 - Read the text by	ATV2 -

M5 08/09 CH=8 H	Distance Learning?	learning	distance learning from conventional education	Jose Manoel Morran.	Participate in a debate in the chat room.
A2- M5 14/09 CH= 6H	Theories and Philosophies of Distance Education	- Learning theories in distance learning	Know the theories behind distance learning	ATV3 - Use the flash presentation on learning theories.	ATV4 - Discuss the best theory to apply to education, with new technologies.

Chart 2 - Activity Map Source: Franco, 2007.

In this reading, each column meets a planning requirement: the -Lesson" column (dates and workload) indicates how the course is divided; the -Main Theme" column gives the title of the most important content of the week; the "Sub-Themes" column is the subdivision of the main theme into smaller units; under 'Specific Objectives" the pedagogical goals for the module should be entered; under "Activities", the tasks that the students will have to carry out during the module are listed; the columns -Theoretical activities and distance learning resources/tools" and -Practical activities and distance learning resources/tools" indicate whether each proposed task is practical or theoretical and also indicate which environment resources will be used to implement the tasks.

According to Costa (2012, p. 7), the Activity Map allows the instructional designer to organize their work in an integral and complete way, including, among other possibilities:

- distribute the workload between classes;

- establish themes and sub-themes for each lesson;

- define the specific objectives for each lesson;

- establish types of activities (theoretical and practical), their tools and resources;

- define the forms and types of evaluation, as well as the timetable for carrying out this activity;

- plan specific lessons and activities to be taught in a virtual classroom;

- define how the content will be presented and develop the defined materials;

- plan dynamic virtual group or individual activities; and

- plan activities with asynchronous interactive events, such as discussion forums, and

29

synchronous ones, such as chats.

As a complementary resource, Filatro (2008, p. 62) presents a *storyboard* that follows the complete model, a resource that complements the Activity Map in the form of a graphic sketch and showing the sequence of activities that the students will go through in each class to carry out the requested tasks, guiding the production team (Figure 2).

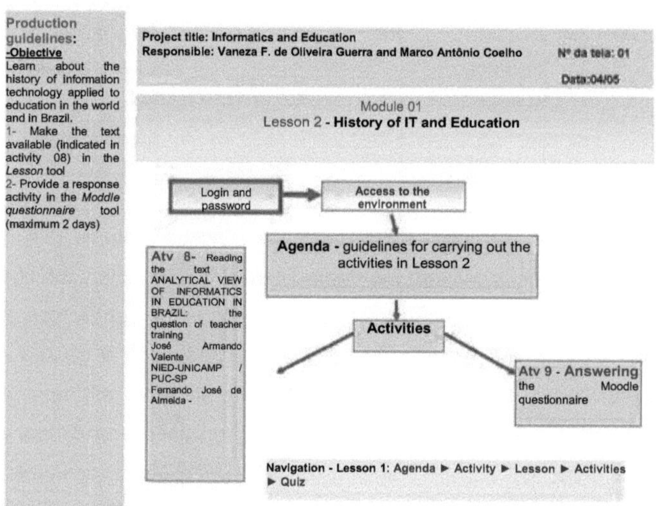

Figure 2 - *Storyboard*

Source: Franco, 2007.

According to Filatro (2008), the *storyboard* has several functionalities. Among them, he highlights

•   specify/detail the content to be presented both in text format and in image and sound format;

•   provide guidance on the activities to be carried out during the course;

•   demonstrate the dialogues of the characters by creating scenarios for them (when necessary), as well as the *voiceovers*; and

•   determine the sequence in which the dialogues and voice-overs will appear in the units.

To increase the level of detail and consequently reduce the complexity of the Activity Map and complement the planning, Filatro (2008) indicates the DI Matrix. Costa (2012) explains that the level of complexity from a pedagogical point of view cannot be very detailed in the

Activity Map, as it requires more detail, and adds that the DI Matrix is a spreadsheet in which the guidelines and procedures for the most complex activities in a course are more detailed.

Dynamic description/proposal	Objective(s)	Criteria / evaluation	Type of interaction	Deadline	Tool	Supporting and complementary content	Student production / evaluation
Based on the text read, draw up a concept map. Reading the text "ANALYTICAL VIEW OF INFORMATICS IN EDUCATION IN BRAZIL: the question of teacher training José Armando Valente NIED-UNICAMP / PUC-SP Fernando José de Almeida	Understanding concepts by establishing links between them	Creativity, good co-relation between concepts	Individual	2 days	Tasks	Text in the L^ăo tool	Text produced in a Word document
A student starts writing a text on a certain subject and passes it on to the next student in the discussion forum.	*Establish coherence and textual cohesion in ideas with the group. *Interaction *Getting to know the other students	Coherence, textual cohesion, interactivity	Group	4 days	Discussion forum	Text on Modlle's own platform	The text must be produced on the Modlle platform itself

Chart 3 - DI matrix Source: Franco, 2007.

This is a DI resource and describes the main dynamic activities of the proposed course. It can be seen that Table 3 highlights the proposal of the dynamic, its objectives and the criteria used in the evaluation, as well as stipulating the deadlines and the tools used. It is an important tool, as it provides the information needed to synchronize the development and implementation team (FRANCO, 2008).

Filatro (2008, p. 44) explains that

[...] using the matrix, we can define which activities will be necessary to achieve the objectives, as well as list which content and tools will be needed to carry out the activities. [...] The matrix also allows us to check the levels of interaction between the student and the content, the tools, the educator and the other students and what kind of virtual environment will be needed to carry out the activities.

It is therefore understood that the DI plays a fundamental role in learning based on the use of communication and information technologies, which expands the structure of planning, organizing and implementing courses in the form of distance education.

However, there is still a lot to uncover, especially since *online* initiatives - computer-mediated distance learning, distance education, open learning, cooperative networking of knowledge and information - are only just beginning.

2.2 Andragogical practice in *online* education

As far as distance learning is concerned, it can be seen from its self-directed learning postulates that there is no specific theoretical concept to support its modeling; however, it can also be seen that the greatest demand for this modality suggests that it is made up of an active and autonomous individual, in other words, an adult. It is therefore necessary to investigate what would be most appropriate in terms of learning orientation for this audience.

In both face-to-face and *online* education, it is necessary to consider the specificities of students, most of whom are considered adults. These individuals, with their particular learning characteristics, have characteristics that pedagogy does not take into account: the way adults learn is one of these factors. The fact is that we are not always faced with a situation in which pedagogy is the best way to conduct learning. Andragogy is therefore a field of research that needs a differentiated approach, as it arose as a result of a new vision of learning.

Andragogy, as a new field of study, questions the model of pedagogy used in adult education, which for Oliveira (2009) is a practice as old as the history of the human race, although it has only recently been the subject of scientific research.

Adults are defined by Belan (2005) as individuals who are mature enough to take responsibility for their actions in society, because they are aware of their actions and are capable of making responsible decisions in their lives. Their desire to learn is directly proportional to their needs and experiences, as they want to apply what they learn immediately to their daily practice.

The concept of adult defended by Oliveira (1999) is that of an individual who is mature enough to take responsibility for their actions in society. It is assumed that this individual, unlike the one Pedagogy focuses on, is capable of making choices that affect their line of production and their interaction with professional experiences, as they have cultural baggage.

Oliveira (2009, p. 1) states that:

Confucius and Lao Tse in China; Aristotle, Socrates and Plato in ancient Greece; Cicero, Evelid and Quintillian in ancient Rome, were also exclusive educators of adults. These great thinkers' perception of learning was that it is a process of active inquiry and not passive reception of transmitted content. That's why their educational techniques challenged the learner to ask questions.

For Filatro (2003, p. 94) -the term Andragogy was originally formulated by the German professor Alexander Kapp in 1833". This context is confirmed by Vogt and Alves (2005, p. 14) who also state that -Andragogy, as a theory or system of ideas, concepts and approaches to adult learning, was disseminated in the United States by Malcolm Knowles during the second half of the last century". His postulates are an important reference on the subject, so much so that the term Andragogy and the name Knowles have become strongly linked.

In this modality, the teacher's main role has changed: he or she is no longer the holder of knowledge and organizer of information, but decides on the various aspects of teaching, planning, content, assessment and methodology. As Belan (2010) shows.

But when teachers decide to articulate their learning roles through the lens of andragogy, their stance needs to be revised and transformed into a facilitating agent with a vision of the different forms of learning, as Belan (2010) describes. This new teacher/facilitator must present information and techniques through teaching techniques and create a suitable environment for learning (p. 55).

The view that the andragogical model is applied as a business education technique is not shared by Belan (2010), as at no point does he admit this possibility. However, a critical reflection on Andragogy applied to conventional educational institutions shows that this model has been used more frequently in companies than in conventional education. At this point, Andragogy tends towards utilitarianism and can often be used by students in a practical way, usually in the workplace.

By linking technology to the industrial development of a globalized culture, which began to be required as specialized training so that workers could meet the demands of the socio-economic moment, we are faced with new business needs, including in the educational sphere. As a result, a type of student has emerged who has more specific needs and who is not in an academic environment, but is part of a corporate structure. This type of student focuses their experiences on their field of work, with a view to gaining their own skills.

Tennant (1997) argues that the value system of Andragogy is centered on the personal and puts the group of adults in the background. Hartree (1984) questions whether Andragogy is a theory or just a set of assumptions about learning.

According to Oliveira (2011, p. 4), Andragogy is a science that is not widely disseminated in Brazil and that the principles proposed by Malcolm Knowles, despite being criticized, are seen as a possibility for rethinking the role of the teacher". It is believed that this theory has a lot to contribute to education, as it enables the educator to seek an action more suited to the adult category.

However, it is accepted that no theory is complete in itself, because in educational practice, most of the time, there is a mixture of concepts, which is thought to be the most appropriate for the teaching objective. But it should be recognized that despite the utilitarian nature of Andragogy, any educational model, as long as its effects have been tested, can be used as a method and applied at any educational level.

2.3 Heutagogy: self-directed learning

Belan (2008) believes that educational and corporate institutions need (and sometimes demand) fast and flexible learning that allows them to develop their employees' capacity to learn for themselves and improve their personal skills. In this context, distance learning is a viable alternative, due to its flexibility in terms of space, time and the diversity of its learners' profiles.

In such a diverse environment, it is important to consider the analogies and differences between the learning of adults (Andragogy), children (Pedagogy) and groups with more specific needs, defended by Heutagogy, as highlighted by Almeida (2008, p. 105):

The concept of heutagogy (*heuta* - self, own - and *agogus* - guide) arose from the study of self-learning from the perspective of shared knowledge. It is a concept that expands the concept of andragogy by recognizing everyday experiences as a source of knowledge and incorporates the self-direction of learning with a focus on experiences.

Almeida (2008) also states that Heutagogy involves the study of self-learning; it values everyday experiences as a source of knowledge and incorporates self-directed learning, focusing on experiences; it tries to explain these phenomena, from the perspective of shared knowledge, which can occur in distance learning. This concept arises at a time when available information needs to be processed and selected, and thus transformed into knowledge. Basically, Heutagogy considers the fact that students manage their own learning flexibly, defining the forms and behavioral models that facilitate their search for knowledge.

In these times of great access to information, Heutagogy doesn't deal directly with the teaching-learning relationship, it goes deeper into the discussion of learning. The issue is therefore one of individual development. How do we learn to learn? The proposal is that content and delivery models should be designed and prepared for the ability to learn the process of acquiring knowledge (HASE; KENYON, 2000, p. 2).

For Batista (2008, p. 32), Heutagogy is the method by which the student defines "[...] what, how and when to learn"; it also signals that the student is responsible for learning, in line with technological innovations, and he adds:

It's directed study, self-learning through practical experience and, once in a safe environment, the more you make mistakes, the more you learn. Through technology, students can not only define "how", but also "when and where to learn".

By autonomous learning, Knowles (1990) means a teaching and learning process centered on the student, whose experiences are used as a mechanism that makes the teacher become the learner's resource.

In order to use the heutagogical, self-directed approach, there is a need to make the teaching-learning process more flexible, in which the teacher is responsible for providing the resources, but the learner is the one who designs and outlines their path, according to a system of negotiations.

Ventura (2001; 2004) and Lebel (year *apud* VENTURA, 2001, p. 35) explain that in the preliminary phases of educational proposals, the negotiations that can be made are: - consultation, proposition, counter-proposition, argumentation, evaluation and attempted conclusion". These negotiations can be understood as contact, agreement to discuss together, knowledge of the object to be negotiated, argumentation and decision.

By observing these stages of negotiation, learners can read about critical aspects or issues and determine what is of interest and relevance to them and, from there, negotiate further reading and tasks (HASE; KENYON, 2000).

Belan (2008) makes a connection between teaching models and the choice of content and methods, and the participation of the teacher and student in this context.

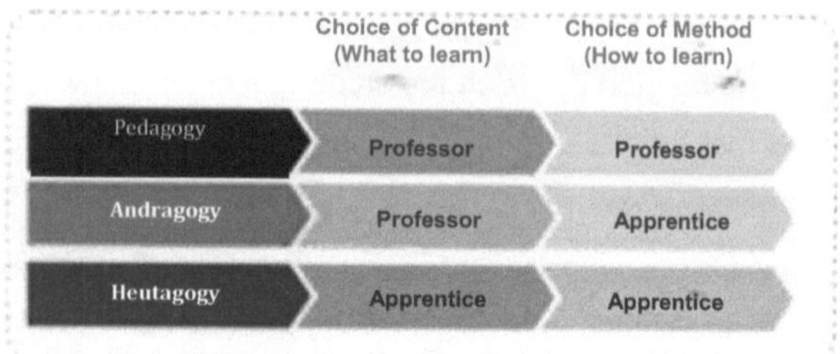

Figure 3 - Comparative table of teaching models Source: Belan, 2008, p. 18.

In Figure 3, Belan (2008) proposes an analysis of the profile of each subject in relation to the teaching model. It can be seen that in the predominantly pedagogical environment, the teacher is the main focus in both the choice of content and the teaching method. As they progress to the andragogical environment, the teacher loses ground to the learner in the choice of content to be studied. In the heutagogical environment, the learner assumes a role that is totally independent of the teacher. Belan (2008) states in the course of his work that it is necessary to first survey the profile of the student in this modality, by means of questionnaires and interviews, and suggests applying Gardner's (1983) multiple intelligences test[1] as the guiding principle for a heutagogical approach to *online* teaching.

The author provides some basic concepts and practical exercises to discover styles, how to retain, and defines some learning strategies based on quizzes, games and logic tests, which can lead the student through a process of learning to know and learning to learn.

Moran (2000, p. 79) argues that learning to know implies learning to learn, understanding the learning process as a method that is never finished.

Self-directed learners therefore require differentiated attention from face-to-face learners. A change of profile is needed. In addition, the heutagogical approach emphasizes the provision of diverse sources of consultation, rather than syllabus content. In other words, the student is not obliged to use any of the instructional material provided.

Negotiations then take place to decide on the format of the assessments, which should become learning experiences rather than just an assessment of an objective achieved. If an outcome or assessment is designed in the right way and negotiated and few drivers are provided, learners will have to try to assess the issues, and draw their own conclusions about them (KENYON; HASE, 2001).

---

1 BELLAN, Zezina. Heutagogy - **Learn to Learn More and Better**. Santa Barba d'Oeste: SOCEP Editora, 2008.

In short, Heutagogy imagines an increase in the autonomy of the student, who can choose *what* and *how to* learn in line with their objectives; it consolidates the most recent vision of the success factors in the construction of knowledge in which the instructor or teacher acts as a mediator and guide to the choices of the learners, providing suitable environments and tools for the effectiveness of the process. This gives rise to the phase of directed or self-determined learning, in which each individual ensures and realizes his or her own knowledge-building process.

It is understood that the heutagogical model should be associated with open learning by means of communication technologies, but that it is still a totally new area of study and research, offering scope for further research.

3 NETWORKED KNOWLEDGE: SELECTION, SHARING AND LEARNING

3.1 Distributed knowledge in communication networks

This chapter presents the notions of knowledge networks, how information is selected and shared, highlighting the autonomy in the learning process imposed by the knowledge society. To this end, we draw on studies by Toffler (1990); Lévy (1993); Castells (1999); Moran (2000); Recuero (2000); Siemens (2000); Kenski (2003); Pinto (2007; 2009); Franco (2008); among others.

Changes are taking place in the information society. Moran (2000) warns that these changes are in the ways of organizing, producing goods, marketing them, having fun, teaching and learning. Smaller and more powerful cell phones, *tablets*, *notebooks* and other electronics, whether they like it or not, are part of the daily lives of children, young people and adults.

These technological tools, when seen as digital educational resources, cause changes in the way distance learning courses are implemented; the speed with which technological transformations are imposed has required new directions in the task of teaching and learning.

Based on the new ways of teaching, with the use of ICTs in the teaching process, new needs and new theories have also emerged, allowing us to better understand these paradigms. The pedagogy that structures teaching processes will never become obsolete, but will undergo changes to adapt to the new context.

The role of the teacher and the student in this new environment is much questioned, but perhaps not discussed with enough concern for the new networked models. Education continues to be static, based on patterns that dilate over time without any change. In this context, we are dealing here with the new heutagogical, shared, hybrid and connected educational models of the knowledge society, which will provide support for a more detailed study of open networked learning and for understanding its particularities.

As Moran (2000, p. 11) shows:

Many ways of teaching today are no longer justified. We waste too much time, learn too little and are continually discouraged. Both teachers and students have the distinct feeling that many conventional lessons are outdated. But where to change? How can we teach and learn in a more connected society?

It's called the "knowledge society", according to Toffler (1990). At the turn of the millennium, educational issues will necessarily require the integration of NICTs, not just as a means of improving the efficiency of systems, but above all as pedagogical tools effectively at the service of the formation of the autonomous individual (BELLONI, 2008).

Burch (2012) reports that the notion of "*knowledge* society" emerged at the end of the 90s. The term is used particularly in academic circles and many suggest the terminology "information society".

Given that the concept of "knowledge society" is very broad, Squirra (2005) states that this term has been coined and implemented in recent times in the different scenarios of globalized culture and that they bear the stamp of the modern and dynamic action practices of the most advanced countries.

Bernheim (2008, p. 7) adds:

One of the characteristics of contemporary society is the central role of knowledge in production processes, to the point where the most frequent term used today is the knowledge society. We are witnessing the emergence of a new economic and productive paradigm in which the most important factor is no longer the availability of capital, labor, raw materials or energy, but the intensive use of knowledge and information.

The above quotes highlight the reality of contemporary society, which increasingly demands continuing education from individuals, constant specialization and, for this, autonomy in their studies. In this way, forming autonomous individuals becomes one of the great challenges for education, since it involves learning to teach and to learn, integrating face-to-face, hybrid, distance and networked studies.

The research of Professor Augusto de Franco (2008) is used here as a basis for initial studies on networks, describing the networks referred to in a very simple way. Based on Paul Baran's diagrams, the professor explains the origin and development of communication networks and their metamorphosis over time.

To better explain the notion of digital networks, Franco (2008) turns to Paul Baran (1964), creator of the first network topology, which would later become the Internet we know today. Baran instituted the distributed network model in order to create more robust communication networks. At the time, the existing models were centralized and decentralized. These models, created with military pretensions for communication, didn't serve their purpose, since their topology didn't offer security if one of their centers were disabled. This would result in communication being cut off with all the other nodes on the network. This is not the case with a distributed network, because even if some nodes go down, the others remain connected, thus not compromising communication between one node and another.

Figure 4 - Network diagrams

Source: Diagrams by Paul Baran (1964), improved by Rodrigo Araya and published by David de Ugarte in the book -The Power of Networks" (2008).

According to Franco (2008), in the three drawings in Figure 4, the points (nodes) are the same, what varies is how they are connected. Networks proper are only distributed networks (the third graph). The other two topologies - centralized and decentralized - can be called networks, but only as particular cases (in mathematical terms). Both are actually hierarchies.

Franco (2008) also shows that for networks to be articulated, it is necessary to connect people or networks themselves, distributed networks. By connecting people with equal interests, group activities and student/student, student/teacher, student/institution and institution/institution relationships are enhanced.

Castells (1999, p. 498) also contributes by stating that the concept of a network is based on a fairly simple definition: "a network is a set of interconnected nodes". And so he postulates:

[...] networks are open structures capable of unlimited expansion, integrating new nodes as long as they can communicate within the network, i.e. as long as they share the same communication codes (e.g. values or performance objectives). A social structure based on networks is a highly dynamic open system susceptible to innovation without threatening its equilibrium (idem, p. 499).

This definition shows that nowadays the change in the form of social organization is independent of the centralized and dominant power relations of the information age, indicating that "[...] the new economy is organized around global networks of capital, management and information" (idem, p. 499), and that "[...] the processes of social transformation synthesized in the ideal type of network society go beyond the sphere of social relations and production techniques: they affect the sphere of social relations and production techniques: they affect the network society. 499), and that "[...] the processes of social transformation summarized in the ideal type of network society go beyond the sphere of social relations and production techniques: they affect culture and power in a profound way" (idem, p. 504).

Recuero (2000) identifies the first element of social networks as the actors, represented by the nodes. These are the people involved in the network being analyzed. As parts of the system, actors act to shape social structures through interaction and the formation of social ties. The actors, as the author refers to, are the users.

Barnes (1987, p. 163) conceptualizes networks by citing the social sciences and states that

[...] a network is the set of social relationships between a group of actors and also between the actors themselves. It also refers to movements that are not very institutionalized, bringing together individuals or groups in an association whose boundaries are variable and subject to reinterpretation. For anthropologists, the notion of social networks seeks to support "the analysis and description of those social processes that involve connections that cross the boundaries of groups and categories".

For Lévy (1993), the social network is based on the permanent exchange of information, stating that if every process is an interface, and therefore a translation, it is because no message is transmitted as it would be in a neutral medium, but rather must overcome discontinuities that metamorphose it.

The interaction provided by communication networks can apply the spaces of communication with other knowledge. This statement by Kenski (2003), who also defines connection as a -moment in which many come together around the same idea", refers to one of the principles of Connectivism, which defines learning as -a process of connecting specialized nodes or sources of information".

Kenski (2003) also states that connectivity occurs when two or more people come together mentally, interact, talk or collaborate. With the help of telegraphs, radios, telephones or digital communication networks, these people can be in different, distant places.

The message itself is a moving discontinuity on a channel and its effect will be to produce other differences, thus working with the notion that the network means working in conjunction with the idea of information (LÉVY, 2006).

Networked knowledge has a point of convergence in interaction and cooperation between peers. When circulating in communication networks, the individual must filter and interpret the information, which is normally laden with subjectivity and comes from an actor who shares and distributes information in the form of knowledge with their intellectual and cultural background. It is this information and its sharing that is the focus of the study of knowledge networks, and it is through them that individual knowledge can be the guide for partnerships that bring reciprocal benefits.

In order to better understand these knowledge networks and get a clearer picture of what they are, Creech and Willard (2001, p. 9) describe some of their advantages:

- Knowledge networks emphasize the creation of common values by all their members, moving through the sharing of information, with the aim of gathering and creating new knowledge;

- Knowledge networks strengthen the research and communication capacity of all members of the network;

- Knowledge networks identify and implement strategies that require greater commitment from those responsible for decision-making, because they move knowledge around within the policies and practices adopted by the participants.

It is thus understood that networked knowledge can develop differentiated learning capacities, producing significant knowledge in a cooperative format.

When approaching the term network in a generalized way, it is understood that it is impregnated with the concept of interaction and, as a consequence, the sharing of information. In shared information networks, the individual occupies the space of a node and can share knowledge with other nodes in the same or other networks.

Guimarâes et al. (2003) state that the insertion of the individual into a knowledge and information network occurs through the establishment of relationships and has as its central focus the cooperation that sustains the network and promotes sustainable development, expanding it.

Tamael (2005) also contributes by highlighting the benefits of networking: building knowledge; technological development; increasing the quality and productivity of services, products and processes.

It is therefore understood that information and knowledge networks are recent instruments of the information age and require a different look at their possibilities for application in education, and should be the subject of further research so that we can delve deeper and move in new directions.

It is also undeniable that society is constantly changing, and that it will be necessary to review concepts and change attitudes at certain levels, institutional, financial and property, in order to adapt to the new molds. At the same time, education will have to find new ways of working and take advantage of this new concept of networks. New methods are needed to create new directions in the search for educational quality.

1.2 The *e-teacher* and the e-student in the perimeters of emergent education

The discussion of teaching attitudes, of the qualities of teachers' professional practice in the light of the demands of educational work, needs to become a relevant element for educational change these days. This attitude reveals the urgent need for a review of training models in teacher development, which implies recognizing this attitude as a set of behaviours, skills, competences and values that should constitute the role of the teacher. It is significant, therefore, to think of training as a permanent and dynamic process based on the theory-practice relationship and the different situations of everyday life in the classroom (BRITO, 2006).

Modernity requires teachers to undergo not only initial training, but also ongoing continuing training, so that they can provide a foundation for their work that meets the needs of their job.

Teaching, favored by the distance learning modality, has demanded a closer look at teaching/learning methodologies and mechanisms. In contemporary times, teachers must be prepared to understand the new forms of social interaction brought about by digital technologies - new spaces and different methodologies for developing skills and competences.

Consequently, there is a need for changes in teacher training. To this end, there is a need for new learning resources in the construction of knowledge, due to the demands that information and communication have made in the context of distance learning.

In this way, the changes must be both spatial and in relation to the resources and strategies for building knowledge, due to new demands for information and communication that are gaining ground beyond the culture of paper, in other words, the use of digital space.

Lévy (1998, p. 28) states that

[...] the construction of knowledge is now equally attributed to the groups that interact in the space of knowledge. No one is in possession of knowledge, people always know something, which makes them important when they work together to create collective intelligence. "It is an intelligence distributed everywhere, incessantly valued, coordinated in real time, which results in an effective mobilization of competences."

It is worth noting that digital culture enables more dynamic and faster training. What's more, the structure and functionality of technology meet the training needs imposed by contemporary times. This training ranges from a one-off piece of information, such as consulting a traditional encyclopedia, to various courses at different levels that can be taken away from the location of the institution promoting it, through distance learning.

When it comes to distance education, new media allow innovative ways of producing, editing, publishing and interacting. In this sense, Soares (2002, p. 151) refers to the computer screen as a new writing space capable of "[...] producing significant changes in the process of interaction between writer and reader, between writer and text, between reader and text, and even, more broadly, between human beings and knowledge".

In this emerging space, individual roles are changing with the use of NICTs; the teacher is no longer the holder of knowledge and is now considered a mentor. The vast amount of information and social movements moving towards the age of knowledge and information suggest the need for this change. In this sense, the figure of the teacher/tutor emerges. So who would this person be: a teacher or a mere actor in the process?

According to Baptista (2011), the etymology of the word tutor reflects what its functions should be. It derives from the Latin *tutororis*, the one who has guardianship over someone, the protector and defender; and has its root in the verb *tuéor* which means to see, observe, discover, protect and defend.

According to Litwin (2001, p. 93), "a tutor can be defined as someone who guides, protects or defends someone in any way, while a teacher is someone who teaches something".

Brzezinski (2012) also agrees with these definitions:

The field of education began to use this term from law, when it was transported to the academic field by the English, who began to practice tutoring and conceptualized this practice as the individual and close relationship established between the teacher and the learner. This relationship is established in an environment marked by respect and friendship, mediated by the partnership between teacher and student, with the aim of continually stimulating the student's curiosity. (p. 125)

Sà (1998) reports that tutoring as an artifice was born in the 15th century at university, where it was used as religious guidance for students, with the aim of instilling faith and moral conduct. Later, in the 20th century, the tutor took on the role of advisor and accompanier of academic work, and it is with this same sense that it has been incorporated into current distance education programs.

But what are the tutor's duties? According to Resolution

CD/FND No. 26, of June 5, 2009, which establishes guidelines and directives for the payment of study and research grants to participants in the preparation and execution of courses in higher education, initial and continuing training programs within the scope of the Open University of Brazil System (UAB), the pertinences to the tutor are:

- Mediate the communication of content between the teacher and the students;

- Monitor student activities, according to the course schedule;

- Supporting the subject teacher in the development of teaching activities;

- Maintain regular access to the Virtual Learning Environment (VLE) and respond to student requests within a maximum of 24 hours;

- Establish permanent contact with students and mediate student activities;

- Collaborate with the course coordinator in student assessment;

- Participate in training and updating activities promoted by the educational institution;

- Draw up monthly student monitoring reports and send them to the tutoring coordinator;

- Participate in the evaluation process of the subject under the guidance of the teacher responsible; and

- Provide operational support to the course coordinator in face-to-face activities at the centers, especially in the application of assessments (BRASIL, 2009b, p. 3).

These duties shape the tutor's profile and at the same time contribute to shaping their identity. Brzezinski (2012 *apud* Almeida, 2012) infers that the tutor's identity as a teacher is shaped by their teacher training and the course in which they carry out their tasks; by their skills in integrating the tasks made dynamic by communication flows; by the mediation between the student and the knowledge brought about by intervening in the learning process; by the interpersonal skills that make up the affective aspect of the teaching relationship; and by their technological knowledge.

The MEC defines two roles for teachers in the distance education system who are tutors: "face-to-face tutor" and "distance tutor" (BRASIL, 2007, p. 21). The face-to-face tutor is the professional who attends to the student in a real location, school or institution, guiding them through their activities, helping them to coordinate their time and studies.

The face-to-face tutor is the most immediate figure for the students and the relationship between them must be structured with a considerable degree of affection. The MEC suggests one face-to-face tutor for 25 students, and it is considered that these proportions meet the quality requirements.

The distance tutor is responsible for mediating and accompanying the student, offering support in relation to the content taught in the subject or course. -The main task of this professional is to clarify doubts through discussion forums on the Internet, by telephone, participation in videoconferences, among others" (BRASIL, 2007, p. 21).

According to Belloni (2006), this professional is also responsible for promoting spaces for collective construction, selecting material and sending material to support the content studied.

Tutors play a fundamentally important role in the educational process of distance learning courses. All studies on distance education agree on the importance of the role of tutors in the success of learning and in keeping students on courses.

One advantage of distance learning is that the student and teacher-tutor don't need to be in the same place for there to be communication between them. In the case of contact by e-mail/discussion list and/or forum, they don't even need to be connected at the same time.

Moran (2009) also states that in a few years' time there will hardly be any totally face-to-face courses. This is why there is a need to look for new tools and methods to strengthen teaching practice and knowledge. It's worth innovating, testing and experimenting, because then progress will come more quickly in the search for these new models that are in line with the demanding changes that have been experienced in all fields, when observing the need to learn continuously.

All universities and educational organizations, at all levels, need to experiment with how to integrate face-to-face and virtual learning, guaranteeing meaningful learning. We need to experience a new pedagogy of communication and management of face-to-face and virtual learning. It's important that universities' distance education centers come out of their isolation and get closer to departments and groups of teachers interested in making their classes more flexible, making it easier to move between face-to-face and virtual (p. 5).

In view of these findings, it is necessary and urgent for teaching to be reviewed in the light of the professional demands - of teachers both in action and in training on degree courses - of a society where information and communication are moving at a rapid pace. In fact, classes need to become more agile and distance learning, through the new information technologies, is becoming a possibility for learning that is necessary to meet the demands of contemporary times. According to Heide and Stilborne (2000, p. 282),

[...] it is, above all, to the qualification of the pedagogical and educational process that distance education makes a fundamental contribution, with the training and updating of education professionals, and with training/specialization in new occupations and professions. Changes, however, do not happen suddenly, which makes a continuous process of reflection, evaluation and research essential if innovations are to take place with greater quality and credibility. In this sense, distance education is a privileged channel for interaction with the manifestations of scientific and technological development. However, it is not technology that will create change in education, but the power of technology that will enable teachers and students to make the necessary changes.

Thus, distance or semi-distance learning, as an alternative to conventional teaching, should enable the acquisition of knowledge by the various teachers in practice and also those in training. However, Mugnol (2009) draws attention to the fact that, although distance education is evolving, some of its main strategic points (objectives, form of transmission, target audience of the courses offered, formation and organization of pedagogical projects, learning assessment methods) have not yet been discussed in the necessary depth; The system for monitoring student learning, teacher training, the different methodologies used and the evaluation of the results of the teaching-learning process are also in need of regulation.

It is therefore necessary to look for possible and viable ways of positively and safely transposing conventional teaching into virtual teaching. To this end, Portal (2001, p. 114) points out that "[...] distance education emerges as an alternative that requires rigorous reflection, especially from educators, to overcome both traditional educational paradigms and the mythification of the technological world".

This new training space/model presided over by virtuality, interactivity and interconnectivity requires more flexible and innovative methodologies, so that learning takes place in a shared dimension. In other words, collaborative learning.

1.3 The role of the student in the context of new technologies

Reflecting on the role of the student in the new model of education that is emerging through NICTs, we soon realize the need to project this actor into the distance learning modality. Virtual teaching is a reality experienced by universities today, and understanding the role of the student in this new educational context is necessary in order to clarify the forms, models and tools to be used in behavioral changes, since this actor is the main objective of all educational research.

The student is a central element in distance and semi-distance learning. In these generative environments, they cease to be mere recipients of information prepared by the educational system and take on the role of researchers. However, Demo (2009) warns that the dominant trend is one of reproduction. Students know far more about copying than producing their own ideas. In this way, they don't find meaning in the information.

According to Moran (2000, p. 73),

Students translate what they have managed to retain or memorize and, over time, this information is forgotten. The teacher often naively believes that teaching is consolidated by the amount of information that is explained in order to be memorized. In turn, students complain that even though they have mastered the information, they are unable to apply it to a concrete situation.

Kenski (1998) warns that the digital style necessarily engenders not only the use of new equipment for the production and apprehension of knowledge, but also new learning behaviors, new rationalities and new perceptual stimuli.

However, as Demo (2009) shows, there are always students who don't like the virtual modality, let alone the internet, mostly because they don't feel comfortable with computers. These people, usually from older generations, attribute their lack of technological fluency to the difficulty of accessing the means of communication.

For Kenski (1998), digital technology breaks with the continuous and sequential narrative of images and written texts and presents itself as a discontinuous phenomenon. In this environment, students become discoverers, transformers and individual producers of knowledge.

Behrens (2000, p. 75) points out that

The quality and relevance of the production also depends on the individual talents of the students, who are now considered to have multiple intelligences. Intelligences that go beyond the linguistic and mathematical reasoning that the school has been offering. As partners, teachers and students unleash a process of cooperative learning in order to produce knowledge.

Thus, for virtual teaching practices to have an effect on students, they need to develop an autonomous attitude throughout the educational process. The fact that activities in this dimension don't take place in a physical space, without institutional timetables, and because there is no teacher present, they need to develop their own ways of organizing time, as well as their own ways of studying.

Therefore, it can be said that both the teacher and the student must adapt to the changes demanded by contemporary society, acquiring a new attitude in order to break with the emerging paradigms involved in the process.

1.4 The semi-presential modality in the new context of higher education

In 2004, by means of Ministerial Order 4.059, the MEC authorized the implementation of semi-presential courses in undergraduate courses at higher education institutions. This decree also mentions that semi-presential courses can only be offered in courses that have already been recognized and that this offer does not exceed 20% (twenty percent) of the total course load, while also stressing that the assessments of these courses are face-to-face.

This same legislation characterizes the semi-presential modality.

[...] as any didactic activities, modules or teaching-learning units centered on self-learning and with the mediation of didactic resources organized in different information supports that use remote communication technologies (BRASIL, 2010b).

According to Moran (2006), semi-presential *learning* (also called *blended learning*) combines face-to-face teaching with the use of distance resources. It can take place in two parts, one in the classroom and the other at a distance, mediated by the use of communication and information technologies (MORAN, 2004).

On the issue of face-to-face contact, Toschi (2008, p. 32) states that

In distance learning, two types of presence can occur: physical presence, which refers to the "real place", and virtual presence, which refers to the "non-real space". Physical face-to-face takes place in the physical dimension of time, in the institutional (collective), chronological and individual aspects. In other words, classes take place in a certain place - institution, school - and at simultaneous times, with several individuals meeting who are biologically present.

Moran (2004) makes a projection of what could happen with the emerging education models when he informs us that with increasingly fast and integrated technologies, the concept of presence and distance as well as the ways of teaching and learning could change. He adds:

We are moving towards an unprecedented rapprochement between face-to-face courses (increasingly semi-presential) and distance learning courses. Face-to-face courses will have subjects that are partly distance learning and others that are totally distance learning. And the same teachers who work in face-to-face-virtual courses will also start working in distance education. We will have countless learning possibilities that combine the best of face-to-face (when possible) with the facilities of virtual (p. 146).

It is possible to see in the author's speech a projection of what could happen in the coming decades with this multiplicity of modalities. The trend is towards merging, which makes us understand that the semi-presential modality is the merging of the distance modality with the face-to-face modality. This idea is clarified by Voigt (2007), who states that semi-presential education is like a bridge linking the classic face-to-face modality with modern distance education, making it possible to enjoy the advantages of both modalities.

In the semi-presential modality, students and teachers are physically separated at certain times during the course, but interconnected through communication, interaction and face-to-face classes. Moran (2009a) corroborates this by saying that this interconnection allows the content studied in non-face-to-face moments to be taken up again in face-to-face classes, stimulating the teaching and learning processes (OLIVEIRA, 2007).

However, Kenski (2003) assures that the virtual educational space expands the face-to-face space, creating new dimensions for access to education, new possibilities for communication and aggregation. The same author also states that regardless of the more or less intensive use of media equipment in classrooms, teachers and students have contact with the most diverse media every day (KENSKI, 2007, p. 85).

It is necessary to emphasize that the media of communication, more specifically, *chats*, forums, videos, hypertexts, *wikis*, *emails*, instant messaging *software* and also telephone, TV and video make up the instrumental apparatus that supports the student in the distance learning modality.

In this sense, Ebert (2003) infers that

The means of communication as promoters of interaction between students and knowledge are another important characteristic of semi-presential teaching. This communication, produced between students and the center producing the course, takes place between teachers, about the work carried out by the students, functioning as a "dynamic and innovative mediation" for evaluating the student's progress. The main medium for this communication is the written word (printed text, books, magazines and virtual text). However, other media are also used, such as telephone, audio, video, software and hypermedia, as well as predefined face-to-face meetings between teachers and students (p.10).

Almeida (2003) also points out that the representation of information in hypertexts using these different media and languages makes it possible to break away from the static and linear sequences of a single path, with a previously fixed beginning, middle and end. In this observation, it is understood that the student does not have a defined path in a virtual learning environment, which allows them to exercise their autonomy by using their learning style to go back and forth through the "sea" of information in search of what is meaningful to them.

The constant demand for updating allows for new productions of knowledge and the breaking of paradigms, and it is based on this context that the semi-presential modality would serve as an additional path, aimed at the student's self-development and self-empowerment.

It is also important to note that "e-teachers" and "e-students" need to be aware that the virtual environment is no better or worse than the traditional environment, but that it is essential to have the vision that one complements the other.

Therefore, when proposing courses in the semi-presential mode, it is essential to understand that for each resource used, it is necessary to know exactly what objective is proposed for its use.

4 FIELD RESEARCH METHODOLOGY

Education is always changing; new texts, new ways of teaching and learning are present in everyday life. Researching new methods, new tools is becoming a commonplace but pressing task in the academic community and according to Fettermann (2012, p. 74),

[...] prompts the search for proposals whose theoretical-methodological approach meets the demands of contemporary times, in which advances in science and technology prevail, amidst the complexity and plurality of codes, languages and languages. In fact, the transformations that are taking place in today's society point to the importance that education, information and knowledge are acquiring, in an overwhelming way never seen in previous eras. Hence the urgent need to improve the quality of education in all its different forms.

In the never-ending quest to improve the quality of teaching mentioned by the author, this chapter outlines the methods used to achieve the proposed objectives in order to obtain answers to the research in this study.

However, before proceeding with the description, it is important to make a few comments about the method adopted to investigate the data. The aim of this study was to analyze the application of connectivist theory in the traditional instructional design of semi-presential courses in undergraduate courses. In order to carry out this analysis, based on the theoretical framework, it was felt necessary to first evaluate the semi-presential modality at Faculdades Vale do Carangola in order to find out what conditions the use of the virtual environment was in and what the behavior of students, teachers and technicians would be.

The authors chosen based on the theme studied allowed for an interdisciplinary dialog between new technologies in teaching and education; places and subjects were connected forming a network in which the following stood out: Skinner (1974), Knowles (1980), Lévy (1996; 2006), Piaget (1997), Hase and Kanyon (2000), Moran (2000), Siemens (2004), Filatro (2008), Kenski (2003), Franco (2008), Belan (2008), among others.

As a first step, authorization to carry out the research was requested through a letter sent to the Pedagogical Directorate and the President of the FAFILE Foundation in Carangola (Appendix 1), so that access could be gained to the distance learning platform, where the semi-presential courses are taught, and to all the documentation, such as the virtual classroom, teacher and student profiles and learning objects.

In this context, there were three categories that made up the Virtual Learning Environment (VLE): teacher, student and support staff. Using a tool for evaluating semi-presential teaching systems developed by Bertolin and Marchi (2010), a questionnaire was applied to the three different categories, with closed questions directed at each category - organized according to a system of indicators, without giving rise to duplicate answers. In this way, 100% of the selected sample of teachers and support staff and 82% of the students enrolled in semi-presential courses in the second semester of 2012 were reached.

4.1 Research universe

4.1.1 The place

According to Carelli (2008), Fundaçâo FAFILE de Carangola, which supports Faculdades Vale do Carangola, is a non-profit organization that invests all its income in maintaining its educational institution, seeking to improve the knowledge and development of both its students and teachers.

In 1990, the Foundation opted to combine its tradition with the strength of the State of Minas Gerais by joining the State University of Minas Gerais, UEMG, which strengthened its commitment to the scientific development of the region.

This partnership has brought the FAFILE Foundation in Carangola not only support in scientific development, but also the advantage of being part of one of the largest higher education institutions in Minas Gerais. Graduates receive the UEMG seal on their degrees, which represents a huge advantage in the job market. In the second semester of 2012, it had 13 ongoing projects, 11 of which were scientific initiation and two extension projects (CARELLI, 2008).

4.1.2 Vale do Carangola Colleges

The Faculdades Vale do Carangola (Figures 5 and 6), a unit maintained by the FAFILE Foundation in Carangola and associated with the State University of Minas Gerais, has 10 degree courses, namely: Degrees in Mathematics, Languages, Biological Sciences, History, Geography, Pedagogy; and Bachelor's Degrees in Tourism, Administration, Social Work and Information Systems. A Bachelor's Degree in Nursing is awaiting approval from the State Education Council. These courses are offered on a face-to-face basis. As of the second semester of 2008, the degree courses have had 20% of their subjects implemented in the semi-presential mode.

In 2012, the MEC gave a grade of three to the project for a fully distance learning theology course and authorized an Open University of Brazil pole at the institution.

Figure 5 - Aerial view of Faculdades Vale do Carangola Source: Braz Cosenza.

Figure 6 - Front view of Faculdades Vale do Carangola Source: Marcos Antônio Coelho.

In the second semester of 2013, Faculdades Vale do Carangola had a teaching staff made up of seven doctors, twenty-six masters, forty specialists and one graduate.

The Vale do Carangola College was chosen as the location for the field research because it has a virtual learning environment (Figure 7) in which all those involved actively participate and attend semi-presential classes.

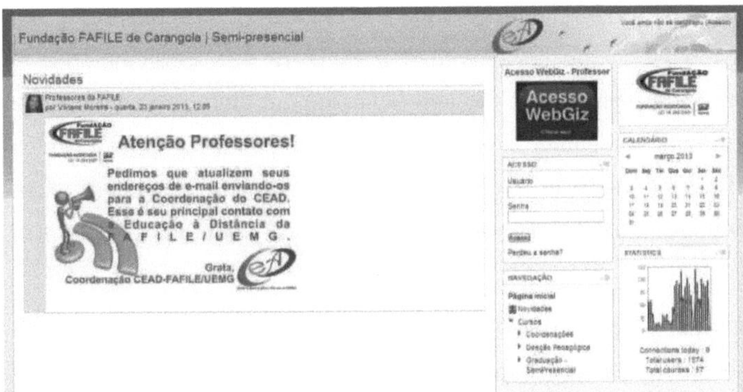

Figure 7 - Initial screen of the AVA of Faculdades Vale do Carangola Available at: <http://www.carangola.br/casp>. Accessed on: January 25, 2013.

4.1.3 Population and sample

This survey has a diverse population separated into three categories: teachers, students and support staff. According to the Coordination of the Distance Learning Center, the number of students, teachers and support staff is as follows: 4 people responsible for technical support, 42 teachers, 176 students from the Administration course, 123 students from the Biology course, 37 from the Geography course, 22 students from the History course, 52 students from the Languages course, 71 students from the Mathematics course and 93 students from the Pedagogy course, making a total of 574 undergraduate students who make use of the semi-presential modality of the virtual learning environment, taking subjects.

The sample defined for the survey was probabilistic, as the *online* questionnaire method made it possible to reach 100% of the defined population.

4.1.4 Type of research

Exploratory research began with a theoretical bibliographic reference which, according to Marconi and Lakatos (1992), is a survey of all the bibliography already published, in the form of books, magazines, single publications and the written press, and currently with material available on the internet.

According to Gil (1999, p. 43), exploratory research "[...] aims to provide an approximate overview of a given fact" and its purpose is to put the researcher in direct contact with the material, helping in the analysis and manipulation of the data collected. It can be considered the first step in scientific research.

Marconi and Lakatos (1996) point out that quantitative research uses interviews, questionnaires, forms etc. as data collection techniques. Gil (1999, p. 128) defines "[...] a questionnaire as a research technique consisting of a more or less large number of questions presented in writing to people, with the aim of finding out their opinions, beliefs, feelings, interests, expectations, situations experienced, etc.". For Marconi and Lakatos (2003, p. 201), "[...] the questionnaire is a data collection instrument consisting of an ordered series of questions that must be answered in writing and without the presence of the interviewer".

A qualitative and quantitative approach was chosen as the structuring basis, with exploratory field research involving data collection using a questionnaire proposed by Bertolin and Marchi (2010), specific for evaluating the semi-presential teaching modality.

Once the type of research, data collection instruments and the theoretical framework had been established, it was possible to look at the authors and demarcate the problem around the proposed objective, thus supporting the investigation.

4.1.5 Data collection instruments

In the hope of clarifying the phenomena and outlining the main guiding aspects of the research, the first technique used was non-participant observation, which, according to Marconi and Lakatos (2010), is the type of observation in which the researcher has contact with the group or reality being studied, without being integrated into it.

In the second stage, we used a questionnaire entitled "Tools for evaluating semipresence courses: a proposal based on indicator systems", by Bertolin and Marchi (2010), adapted to the needs of the moment and sent to the research participants via *Google docs,* in the form of a form.

In order to get a differentiated view of the actors who are part of the semi-presential course system, Bertolin and Marchi (2010) present a list of the roles played by the different subjects as evaluators or self-evaluators for each of the proposed indicators and establish the research subjects: teachers, students and support staff.

4.1.7 Validation of instruments

Bertolin and Marchi (2010, p. 10) state that -In the development of a semi-presential course, the main people involved are the teacher, the students and the staff supporting the distance activities (tutors, designers, etc.)". Therefore, the participation of these subjects is necessary for the assessment of a semi-presential course to be complete and reliable. They add:

[...] specific questionnaires for the different subjects are necessary in order to cover all the aspects (indicators) to be assessed and make it possible to self-assess the assessment itself. This can be done by comparing the responses of the subjects who are self-assessing with the responses of other subjects involved who are assessing a particular aspect. The greater the similarity of responses between self-assessor and assessor, the greater the level of reliability of the assessment carried out. The different subjects play different roles (evaluator or self-assessor) depending on their role as protagonist or user in relation to each indicator (BERTOLIN; MARCHI, 2010, p. 10).

As an argument, Bertolin and Marchi (2010) use the studies of UNESCO's *Latin American* Laboratory for the *Evaluation of Educational Quality* to explain the use of systemic indicators of inputs, processes and results in the evaluation of quality in semi-presential education and report that

[...] the level of quality in education basically consists of defining a set of variables that systematically provide a reliable and valid picture of the state of education systems and that can be used to help guide and improve actions (BERTOLIN; MARCHI, *op cit* UNESCO, 1997, p. 7).

Input indicators aim to assess both the financial, human and technological resources allocated to the education system. Aspects relating to general costs, investments in ICT and the quantity and training of teachers can also be included among input indicators and there is a certain variety of names to represent indicators related to the purposes of education systems, such as "products", "results" or "outputs" of the systems (BERTOLIN, 2007).

The process indicators relate to the primary characteristics, concerning direct participation in the education process, i.e. impact on the pedagogical context; and secondary characteristics, concerning the organization of the primary characteristics. Bertolin (2007) also points out that aspects concerning the number of teaching hours, teacher dedication, access to ICTs and their use can make up the structure of process indicators.

Finally, it can be seen that outcome indicators are directly linked to the results obtained in the education process, evaluations and intermediate purposes; student achievement levels, pass rates and, according to Bertolin (2007 *apud* ESTRADA et al., 1999), student achievement levels in exams, pass rates and schooling rates can be among the outcome indicators.

Thus, Bertolin and Marchi (2010, p. 136) draw up a questionnaire based on input, process and result indicator systems (Table 1).

Table 1 - Basic structure of the system of indicators for higher education

Input aspects	Costs and resources, investment in information technology and the number and training of teachers.
Process aspects	Pedagogical and Organizational Context or Primary Characteristics, relating to direct participation in the education process, and Secondary Characteristics, relating to support for the organization of the Primary Characteristics.
Aspects results	Characteristics relating to the intermediate and ultimate purposes of education. ção.

Source: Bertolin and MARCHI, 2010, p. 135.

For Bertolin and Marchi (2010),

The evaluation experiences linked to evaluation with indicator systems consider that quality in education is a multiple concept that cannot be evaluated by just one isolated aspect and must involve all the fundamental elements of the system or process. Thus, it can be said that it is possible to assess quality in education by making a value judgment on a set of attributes, aspects or indicators about the educational inputs, process and outcomes, or the relationships between them. (p. 134)

Thus, they add responsibility to non-face-to-face activities, stating that it is based, in particular, on six principles: interaction in a collaborative environment, self-learning, flexibility of time and space, the potential use of technological tools, the quality of the methodology, and support.

It can be seen that the development of a semi-presential course involves the teacher, the students and the support staff. And, as seen in previous chapters, the teacher takes on the role of tutor, designer, etc. Therefore, in order for the evaluation of a semi-presential course to be complete and guaranteed, it is suggested that these subjects be involved.

In this sense, in order to reach all the indicator aspects to be evaluated, it is necessary to apply specific questionnaires to the different subjects, which can make it possible to self-assess the evaluation itself. Bertolin and Marchi (2010, p. 140) also state that this can be done by comparing the answers of the subjects who are self-assessing with the answers of other subjects involved who are assessing a certain aspect. In this way, it is possible to assess the similarity of the answers between the self-assessor and the assessor, with a higher level of reliability.

To collect the data, all the students, teachers and support staff selected for the survey were asked to answer the online questionnaire in the computer lab during the second semester of the 2012 academic year. This data will be analyzed in the next chapter of this paper.

5 ANALYSIS AND INTERPRETATION OF RESULTS

The purpose of this chapter is to process and describe the data collected in order to prepare it for analysis of the semi-presential teaching system at Faculdades Vale do Carangola. Based on this analysis, we will connect the ideas proposed in the bibliographical survey and compare the subjects' responses. From there, we will try to find the answer to the problem that guided the research.

The data collection instrument used in the survey is a Likert scale, which establishes a numerical scale in which favorable responses receive the highest value and unfavorable responses receive the lowest values. Bad (1), Bad (2), Fair (3), Good (4), Great (5).

In response to the attempt to assess what the research subjects think of the distance learning infrastructure (learning environment/software, videoconferencing resources, etc.) made available by the institution for the course, the following data is provided.

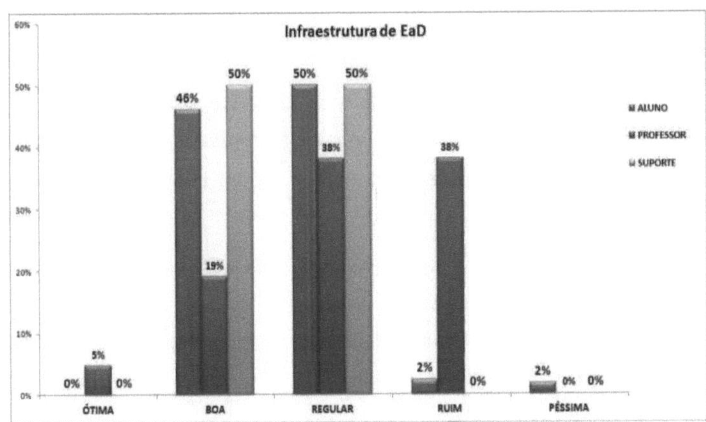

Graph 1 - Distance learning infrastructure Source: research data

Graph 1 clearly shows the comparison between the evaluators of the distance learning structure at Faculdades Vale do Carangola, showing that it is of REGULAR quality, with a relative frequency of 50% of students and support staff and 19% of teachers. 50% of students and support staff consider it to be fair, 38% of teachers with opinions oscillating between REGULAR and BAD.

It is imperative to think of infrastructure as a combination of elements that make up the action of institutional management, and not just as a VLE, although this can be considered the central element.

According to the MEC (2007, p. 24),

In addition to mobilizing human and educational resources, a distance learning course requires material infrastructure that is proportional to the number of students, the technological resources involved and the size of the territory to be covered, which represents a significant investment for the institution.

The MEC (2007) points out that television equipment, videocassettes, audiocassettes, photography, printers, telephone lines, faxes, audiovisual production and videoconferencing equipment, networked and/or *stand alone* computers and others must be used to make up an adequate support structure that meets the quality standards established for distance learning. This does not exempt the institution from having documentation and information centers, such as libraries, video libraries, academic coordination and administrative staff in general.

He also adds that this equipment must be available both at the institution and at the support hubs and also points out that the mobilization of human and educational resources requires a material infrastructure proportional to the number of students and the technological resources involved.

The MEC (2007) also points out that due to the diversity of distance education models adopted, it is essential to have more localized structures; in particular, it suggests that institutions that provide distance or semi-presential education should keep them in their structures:

[The support units for the planning, production and management of distance learning courses, in order to guarantee the standard of quality, need a basic infrastructure consisting at least of an academic secretariat, course coordination rooms, rooms for distance learning tutoring, a library, a teachers' room and a videoconferencing room (optional) (p. 25).

It also recommends that institutions should have certain professionals present in their support units, such as the course coordinator, tutor coordinator, subject coordinator teachers, secretarial assistants and professionals in the various technologies, according to the course proposal (MEC, 2007).

Article 7 of CNE/CP Resolution No. 1/2002 states that institutions must guarantee, with quality and quantity, teaching resources such as libraries, laboratories, video libraries, among others, as well as information and communication technology resources (BRASIL, 2002).

A comparative analysis with the requirements of the MEC shows that there are some gaps in the computer structure of Faculdades Vale do Carangola, which justifies the students' position in their evaluations. FAVALE does not have all the structural technological resources needed to serve the student body. In terms of management, these resources are available, as it has a Distance Learning Center (CEAD), made up of a coordinator and secretariat.

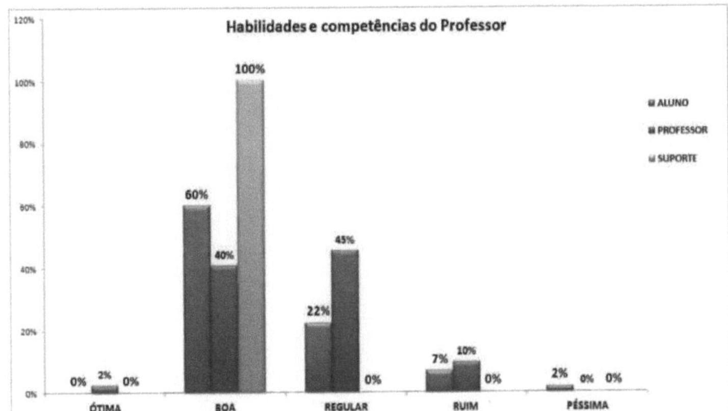

Graph 2 - Assessment of teacher skills and competencies Source: survey data.

Graph 2 shows the question of teachers' skills and competences for working in the semi-presential teaching system. The assessment made by students is GOOD, with 75% of responses. Support staff, 100% of respondents consider them to be GREAT, and 45% of teachers consider their skills and competences to be REGULAR.

Perrenoud (2000) points out that one of the teacher's competencies is knowing how to use new technologies. Competences are inferred to be a set of techniques, skills and attitudes that enable the teacher to play the role of knowledge agent in the semi-presential teaching environment.

The MEC (2000) states that in a higher education institution promoting distance learning courses, teachers should be able to:

a) establish the theoretical foundations of the project;

b) selecting and preparing all the curricular content in conjunction with pedagogical procedures and activities;

c) identify the objectives relating to cognitive competencies, skills and attitudes;

d) define bibliography, videography, iconography, audiography, both basic and complementary;

e) preparing teaching materials for distance learning programs;

f) carrying out the academic management of the teaching-learning process, in particular motivating, guiding, monitoring and evaluating students; and

g) continuously evaluate themselves as professionals participating in the collective of a distance higher education project (p. 21).

FAVALE held regular training courses for staff, tutors and teachers. This justifies the indices shown in Graph 3.

The above demonstration shows that competences are based on mastery and extend to the planning and execution of teaching projects. The opinions on the planning of semi-presential courses are analyzed below.

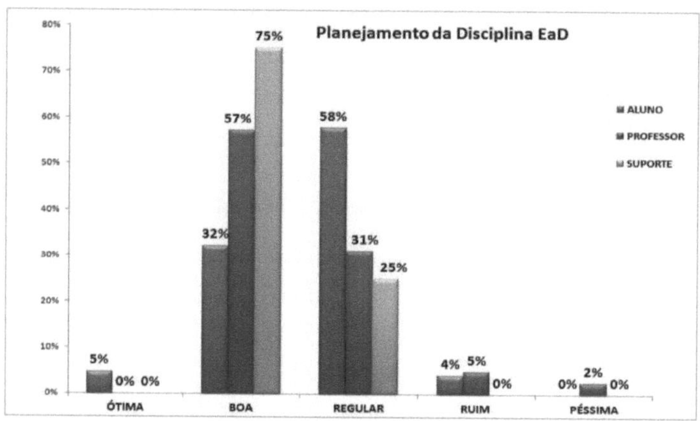

Graph 3 - Distance learning course planning

Source: research data.

According to the results shown in Graph 3, the planning of the Semi-presential course is considered to be of GOOD quality by the self-assessment of the teacher respondents, with 57% and 75% of the support staff. Student respondents, on the other hand, disagree with these answers, with 58% of respondents saying that the quality of planning is REGULAR.

Looking at Graph 3, we turn to Filatro (2003) who sees planning as the main function of the Instructional *Designer* and points to some techniques as principles for developing distance learning that can be of great value to the teacher. These are: developing an Activity Map (Chart 2); DI Matrix (Chart 3) and *storyboard* (Figure 2). These techniques, when implemented, should recognize the learning models, adapting to the subjects and can increase the prospects for learning.

It was also found that the planning done by teachers is conditioned to the face-to-face modality, following the same principles of organization.

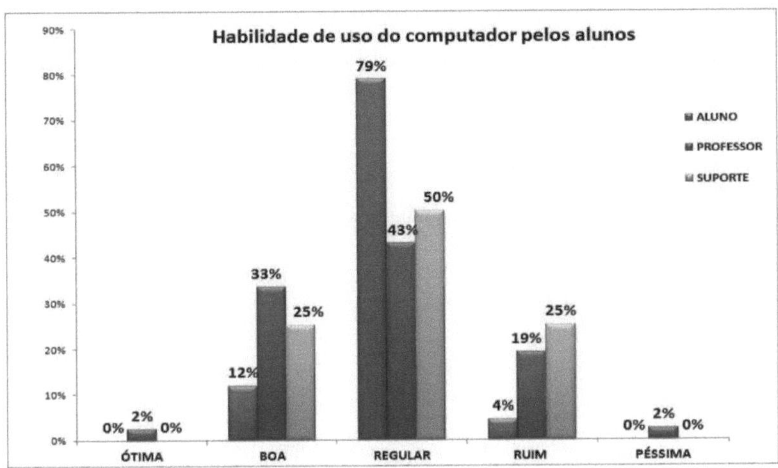

Graph 4 - Students' ability to use computers

Source: research data.

The self-assessment shows that students rate their computer skills as Fair, with 79% of respondents. This view is confirmed by 43% of teachers and 50% of support staff.

The data in Graph 4 shows that teachers admit that they don't train these students before using the subject. Another fact that could help explain this is that according to the socio-economic survey carried out by the college, N% of the students come from rural areas, where there are no computer resources available.

It is essential that training institutions offer physical facilities, human resources and materials that are suitable for the scientific and technological training of future teachers, and when establishing curricular guidelines for the training of basic education teachers at higher education level, they try to solve this gap, demanding that the curricular organization of the courses take into account the preparation of teachers for "[....] the use of information and communication technologies and innovative methodologies, strategies and support materials" (BRASIL, 2002, art. 2, item VI).

From this perspective, it's important to note that educating in this new context isn't just about preparing students to work with technologies and, essentially, the computer, but about seeking new paths that value the construction of knowledge, access to information, freedom of expression and respect for differences.

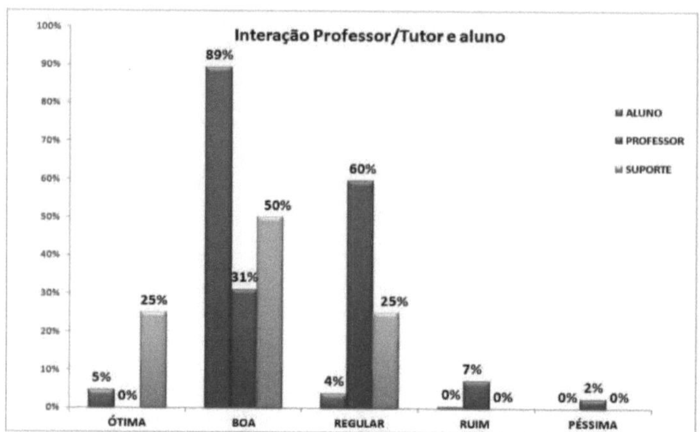

Graph 5 - Interaction between tutor and student Source: research data.

According to Graph 5, the interaction process is considered GOOD by 89% of students, REGULAR by 60% of teachers and 50% of support staff consider it GOOD.

Interaction, understood here as a non-linear communication process, as proposed by Deleuze and Guattari (1995 *apud* KENSKI, 1998), between a sending agent and a receiver, as long as there is an exchange of information through the use of new technologies.

Franco (2010) brings the concept of interaction to digital social networks when he states that interaction is not synonymous with participation and explains why networks are environments for interaction and not participation. According to Franco (2010), participation is designated by a notion constructed outside of interaction and he adds:

To participate is to become part of something that was not invented at the very moment a collective configuration of interactions was established, but something that was (already) given *ex ante*. In an interactive (non-participatory) environment, each person acts on their own terms and not on the terms set by others (as happens in participation). This means, however, that they will be more open or more vulnerable to the unpredictable other. The more interactive an environment is, the fewer anisotropies it causes in the space-time of flows or the fewer deformations it produces in the social field. (p. 3)

Landim (1997) is more specific when he says that interactivity in distance education involves the mediations that constitute the treatment of content and the forms of expression and communicative relationship.

Based on this understanding, it can be seen that the interaction promoted by the teachers in Graph 5 must be based on the use of some kind of communication mechanism and that the teacher must make use of the tools available in order to promote interaction and student participation in the VLE and even outside of it through the creation of knowledge and digital social networks.

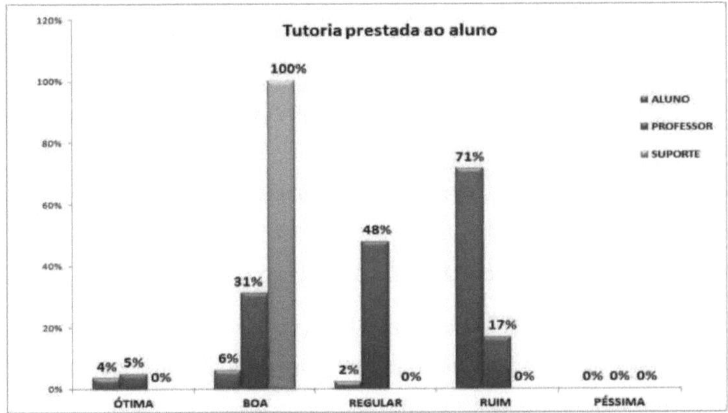

Graph 6 - Tutoring provided to students Source: survey data.

Graph 6 shows that there is a clash of opinions among those interviewed regarding tutoring. In the support staff's self-assessment, which they classify as GOOD with 100% of the responses, the teachers consider it REGULAR with 48%, while 71% of the students classify it as BAD.

In line with what Litwin (2001) suggests, the reality shown in Graph 6 shows that the lack of definition of the tutor's role in this environment can have an effect on student learning, which is clear from the response about the quality of the tutoring system. It is believed that this data reflects a specific reality at the institution, where the role of the tutor is not well defined by management. The tutoring activity is carried out either by the support staff or by the teacher.

This negative effect can be seen in the words of Arnaldo Niskier (1999), when he says that the distance educator

[...] brings together the qualities of a planner, pedagogue, communicator and IT technician. They take part in the production of materials, select the most suitable media for their multiplication, and maintain a permanent evaluation in order to improve the system itself. In this type of teaching, the educator tries to anticipate possible difficulties, trying to get ahead of the students in solving them. (p. 393)

As Niskier (1999) points out, the distance learning teacher must be valued, because his responsibility, in addition to being greater because he reaches an infinitely larger number

65

of students, makes him more vulnerable to criticism and challenges regarding the materials and activities he develops.

Sà (1998, p. 46) emerges from this view by saying that -more is demanded of the tutor than of a hundred conventional teachers, as they need to have excellent academic and personal training".

In addition, the legislation on distance learning does not establish the number of students for each tutor, but by means of quality benchmarks, it does propose that a quantification be made of the ratio of teachers/hours available for the assistance required by the students and the tutor/student ratio. MEC (2007) notes that the appropriate ratio must guarantee good monitoring and communication possibilities between teacher and student.

The MEC's diversity education network[2] , in its proposal for a course for tutors, indicates one tutor for a group of 25 students. We believe that these proportions meet the quality requirements.

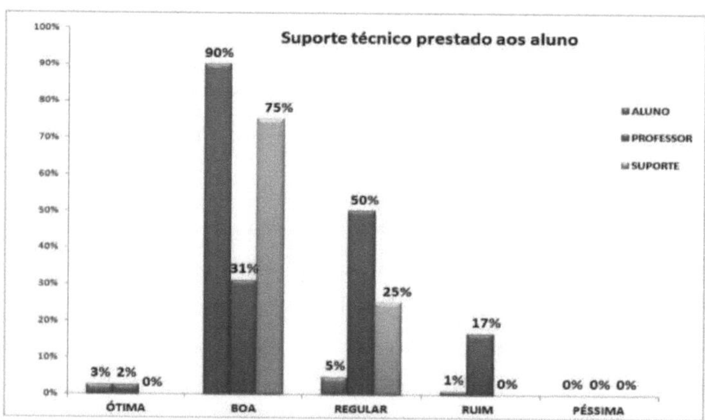

Graph 7 - Technical support provided to students Source: survey data.

The technical support provided to students, based on the self-assessment of the staff responsible, answered the questionnaire with 75% of the responses evaluating it as GOOD. Teachers rated it REGULAR, with 50% of the survey participants, and students rated it GOOD, corresponding to 90% of the respondents, as can be seen in Graph 7.

The MEC (2007) defines technical support as all the professionals who assist the participant in an e-learning or semi-presential course in the technological area, working at

---

2 The Education for Diversity Network (Rede) is a permanent group of higher education institutions dedicated to the initial and continuing training of education professionals. The aim is to disseminate and develop educational methodologies for the inclusion of diversity issues in everyday classroom life. Available at: <http://portal.mec.gov.br/arquivos/redediversidade/>. Accessed on: April 16, 2013.

the face-to-face support centers in technical support activities for laboratories and libraries, as well as in maintenance and janitorial services for materials and technological equipment.

It can therefore be seen that this structure at the Vale do Carangola College still lacks specialized personnel.

It is understood that the student, the teacher, the communication system, the structure and the organization of teaching materials are the main components of an e-learning system, as suggested by Landin (2007). The need to mediate these components is therefore necessary.

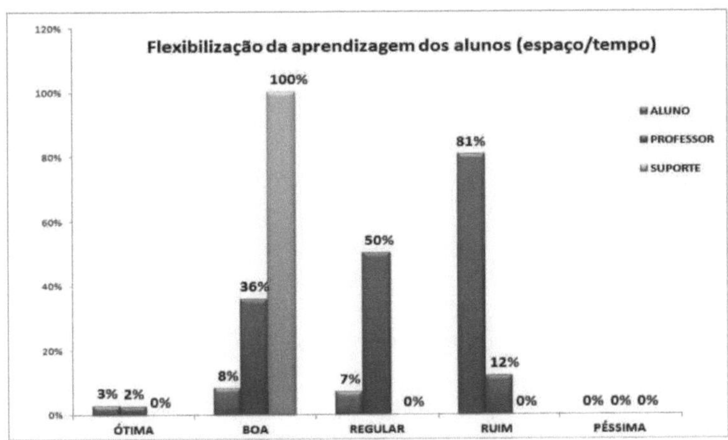

Graph 8 - Level of learning flexibility in relation to time.

Source: research data.

According to the opinion of 100% of the support staff, the level of flexibility in learning, in relation to the time proposed in the semi-presential courses, is considered GOOD, 50% of the teachers rate it as REGULAR and 80% of the students rate this flexibility as BAD.

In order to better understand these results, Bartolin (2007) sought an explanation for the phenomenon. According to Bertolin (2007, p. 136), the flexibilization of time and space:

[...] it makes it possible to carry out the subject's activities at a time and place that is more appropriate for the student, especially those who are already in the job market, reducing the hours spent traveling to the institution and making it possible to carry out the activities according to the needs and characteristics of each student.

It is believed that the modular nature of the VLE, the fact that study time is adjusted, the weekly deadlines for completing activities, and the fact that most of the students at the institutions do not have access to the internet because they live in rural areas, mean that many are unable to keep up with the pace imposed by the content posted by teachers.

It was also found that the media used by teachers does not make learning more flexible, as the tool most often used is text. These factors can influence teacher and student evaluations.

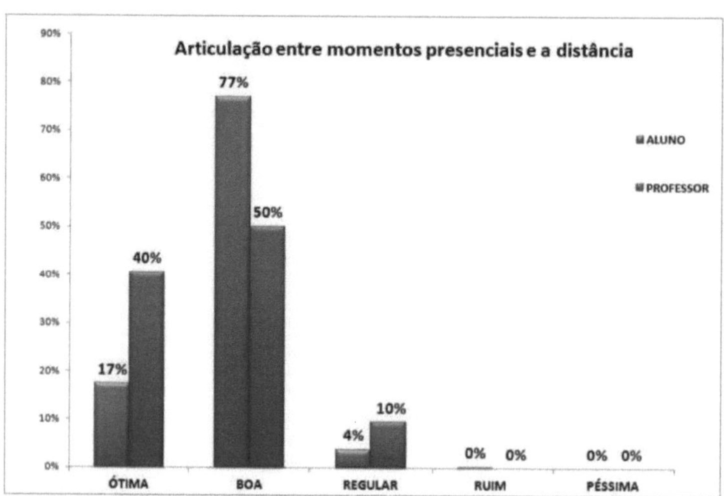

Graph 9 - The articulation and adequacy between the face-to-face and distance learning moments (contents and workload) of the course

Source: research data.

Graph 9 shows that 77% of students and 50% of teachers rate the articulation between face-to-face and distance learning as GOOD, in terms of the content and workload of the semi-presential courses at Faculdades Vale do Carangola. Semi-presential courses work as follows: for three weeks, students attend classes at a distance, and on the last Friday of the month there is a face-to-face meeting, which lasts four hours. It is also important to note that the content to be taught in the semi-presential courses follows the pedagogical project for face-to-face teaching as well as the workload.

This model is criticized by Oliveira (2007, p. 2) when he observes that

[...] a face-to-face meeting is a space for alterity and epiphanic interaction, not a conventional class. Although teachers in the semi-presential modality tend to see it this way, the so-called "face-to-face meeting" is not a class, and it is not where most of the activities of a distance learning course take place.

Oliveira (2007, p. 4) also states that a good sequence for activities in face-to-face meetings would include:

a) beginning with a routine of introducing/integrating the participants, with changes from the first to the last meeting, since it is assumed that personal interactions will deepen throughout the course; b) dialogued presentation or group exercises to resolve doubts and integrate content worked on at a distance; c) group activities to deepen the most relevant aspects, with an exchange of information and points of view between the groups and oral presentation of the results; d) practice of routines that cannot be carried out at a distance; e) assessment activities, either individually or in groups, varying from one meeting to the next to avoid monotony; f) meetings with specialists in the field or with other remote classes, for short lectures, panel discussions, seminars and symposiums, in order to present other views of the content than that of the course designers; g) a final socializing activity among the participants, to ensure motivation to study and improve the interactive tasks carried out at a distance.

These procedures can strengthen relations between students, tutors and the institution, as well as positively influencing the quality of the learning that takes place.

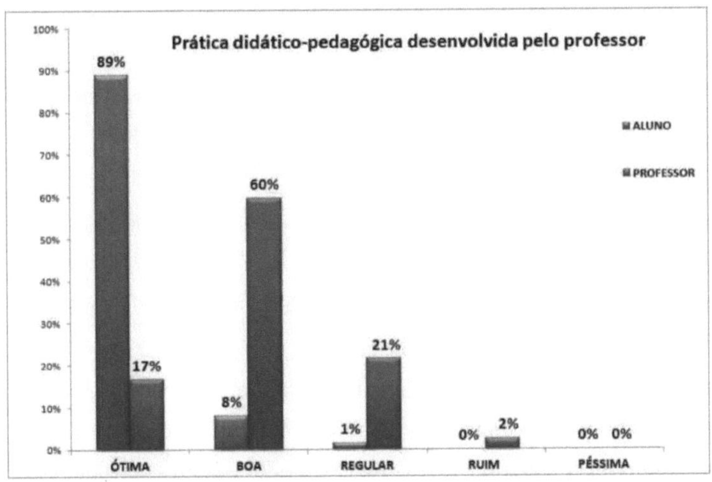

Graph 10 - The didactic-pedagogical practice developed during face-to-face sessions Source: survey data.

According to the information shown in Graph 10, 89% of the students interviewed rated the didactic-pedagogical practice carried out by the teachers during face-to-face lessons as OPTIMAL. 60% of teachers rate their practices as GOOD.

In this context, it can be seen that the pedagogical mediation carried out by the MEC (2007, p.7), from a didactic point of view, "face-to-face moments in distance learning courses are necessary, obligatory and provided for by law". In these moments, students on distance learning courses must attend the institution at predetermined times, to clarify doubts, take assessments or defend final papers. From this perspective, it can be seen that the didactics applied become less important, as the meetings become more informal and interactive, a time when the students get to know each other, ask questions and the teacher remains on the sidelines. The institution will therefore have to provide information on the timetable for face-to-face meetings, in particular the times planned for the course and the strategy to be used.

In this way, it is considered that the planning of a face-to-face meeting should not be structured in the same way as a classroom lesson, nor should it be expected to have the same effects. Oliveira (2007) informs us that the definition of the activities, as well as the time taken to develop them, is influenced by a number of factors such as the number of meetings there will be during the course, the degree of difficulty encountered with the course content, the technological proficiency of the participants and whether or not it is a unidisciplinary meeting.

From another perspective, Demo (2009) recognizes that pedagogy is lagging behind and that it doesn't realize that the world is changing around it. While technology is advancing, pedagogy is lagging behind. He also describes some problems involving teachers when they work in virtual learning environments.

The problems pointed out by Demo (2009, p. 58) include -increased workload without proper remuneration, intellectual property rights and lack of skill with computers, the internet and technology in general"; however, teachers are motivated by intrinsic rewards such as personal satisfaction and flexible working hours.

Referring to teaching skills for working in virtual learning environments, whether in the sense of not knowing how to deal with computers and the internet, or not knowing how to use them for learning, Demo (2009) suggests that institutions offer these technical skills, even though the most convenient way to learn is through practice and interaction.

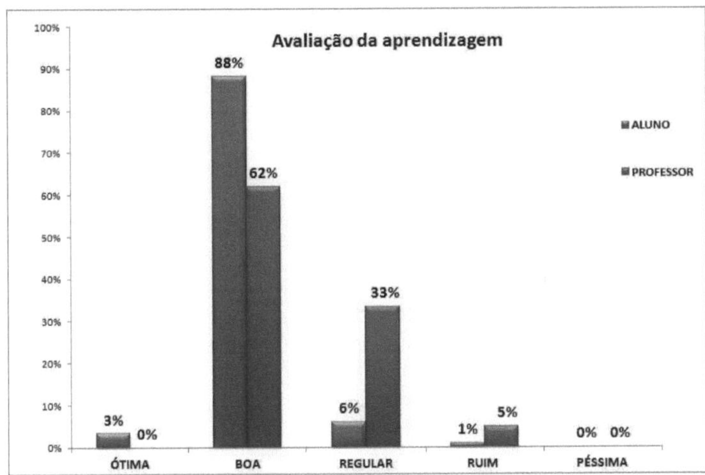

Graph 11 - Evaluation of learning in the course

Source: research data.

Graph 11 shows that 88% of the students and 62% of the teachers rate the assessment of student learning in semipresence courses as GOOD.

The proposal made by Bassani (2006) announces that learning assessment in virtual environments can be understood from three perspectives: assessment through *online* tests; individual student production and analysis of interactions between students, based on messages posted/exchanged through the various communication tools.

In the AVA Moodle (2007) there are some virtual tools such as the creation of questionnaires, multiple choice questions, relationships between columns and automatic evaluation of discursive questions, as well as the evaluation of attendance and monitoring of activities and the individual production of each student.

Moran (2007) warns that, with the Internet, the forms of *online* assessment can be diversified and states that moments of distance assessment can be combined with face-to-face assessment. This makes it possible to group together individual and group activities, and individual and collective construction.

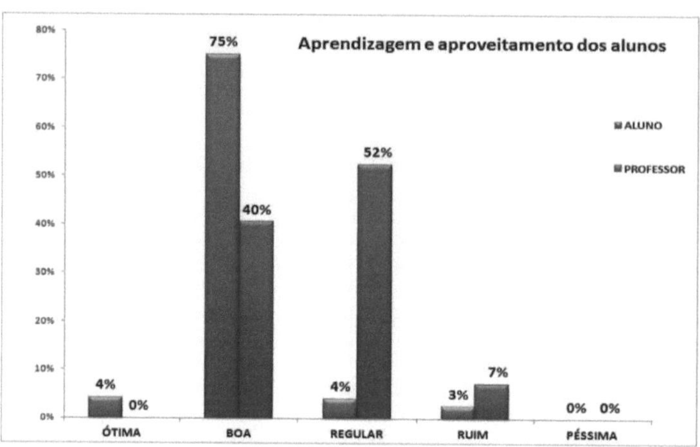

Graph 12 - Student learning and achievement

Source: research data.

Taking into account the student's self-assessment and the teacher's assessment, Graph 12 shows that 75% of the students taking part in the survey rate their performance and learning as GOOD. Teachers, on the other hand, rated 52% as REGULAR.

In order for a learning test to provide meaningful information about student achievement in the semi-presential modality, the teacher must plan it in advance and take into account the expectations and learning worked on during the *online* teaching process.

Rabelo (1998) reports that there is little agreement on evaluation methods, as there are a wide variety of models and, among them, the best way to evaluate. He defines evaluation as

Activity through which, based on certain criteria, relevant information is obtained about a phenomenon, situation, object or person, a judgment is made about the object in question and a series of decisions are made about it (MIRAS; SOLÉ, 1996, p. 375 *apud* RABELO, 1998, p. 69).

When classifying evaluation into categories, Rabelo (1998) explains that most authors agree on these categories, which makes it possible to classify them. With regard to regularity, he states that it is the type of evaluation that takes place continuously, not waiting until the end of a piece of work to carry out an evaluation. As for the evaluator, he explains that it can be internal, when the evaluator who teaches the subject is the one who applies the evaluation, and external, when someone else who has not taken part in the teaching process is the one who applies it (RABELO, 1998, p. 70).

When seen from the category of comparison, Rabelo (1998) divides assessment into normative or criterial. In the first case, it happens when the student's performance is compared with that of other students and in the second case, the student is exposed to the target criteria.

From the training point of view, assessment can be diagnostic, formative or summative. Diagnostic assessment makes a prediction about what the students know about the new content to be covered, giving the teacher the conditions to guide the initial planning and check the relationship between objectives, content and reality.

Through these relationships, formative assessment emerges, which according to Rabelo (1998, p. 72) "[...] aims to provide information about the development of a teaching and learning process, so that the teacher can adjust it to the characteristics of the people it is aimed at". Its main functions include reassuring, supporting, guiding, reinforcing, correcting, etc. It thus assumes a regulatory role in teaching when it offers means of adjustment and strategies to improve the teaching and learning process, checking whether or not the objectives have been achieved.

Rabelo (1998, p. 72) treats summative assessment as a "one-off assessment, which takes place at the end of a teaching unit, a course, a cycle or a two-month period", always trying to determine the degree of mastery of some previously established objectives.

Haydt (1988, p. 18) adds that summative assessment consists of -classifying students according to previously established levels of achievement, usually with a view to their promotion from one grade to another, or from one level to another".

The forms of assessment described by both Haydt (1998) and Rabelo (1998), summarized in Table 4, were designed for face-to-face education. When seen in distance education, it should combine a variety of tools that make it possible to take into account quantitative and qualitative aspects, according to RUSSO (2001).

EVALUATION OF TRAINING				
PERIODS	**TYPES**	**OBJECTIVES**	**INTERESTS**	**SEARCHES**
Home	Diagnostics	Guide Explore Identify Adapt Predict	Student producer	It aims to get to know people's as aptitudes, interests, skills and competencies as a prerequisite for future work.
During	Formative	Regulate Situate Understand	Student activities, production	Seeking information on strategies to solve problems and difficulties

		Harmonize Tranquilize Support Reinforce Correct Facilitate Dialogue	process	that arise
After	Summative	Verify Classify Situate Inform Certify Put to the test	Student as final producer	It seeks to observe global, socially significant behaviors, determine acquired knowledge and, if possible, give a certificate.

Chart 4 - Types of Evaluation Source: Rabelo, 1998, p. 73.

Source: Rabelo, 1998, p. 73.

It can be seen that evaluation seen in phases, presupposing processes of searching for objectives, "gains a mediating dimension in distance education, which projects and envisions the future, supporting an understanding of the limits and possibilities of the students and the permanent adjustment of pedagogical strategies" (HOFFMANN, 2001, p. 59).

Assis (2012), according to Hoffmann (2001), points out that VLE assessment can be understood from three perspectives.

• Assessment through online tests: it is up to the student to answer a set of pre-defined questions and for the computer system to correct them, so that the teacher receives a grade/concept as a final result, emphasizing the product of knowledge;

• Assessment of the students' individual production: in which the final product is overvalued, i.e. the text produced, whether the research is carried out to a certain standard or the questionnaires are solved;

• Analysis of interactions between students: this takes place on the basis of messages posted/exchanged via the various communication tools, with the aim of evaluating the product in the process.

It is important to note that these forms of assessment stimulate students' curiosity and promote autonomy. In this way, e-learning assessments lose the punitive nature imposed by the term "Assessment, Test, Quiz" and become a pleasant tool with educational self-value.

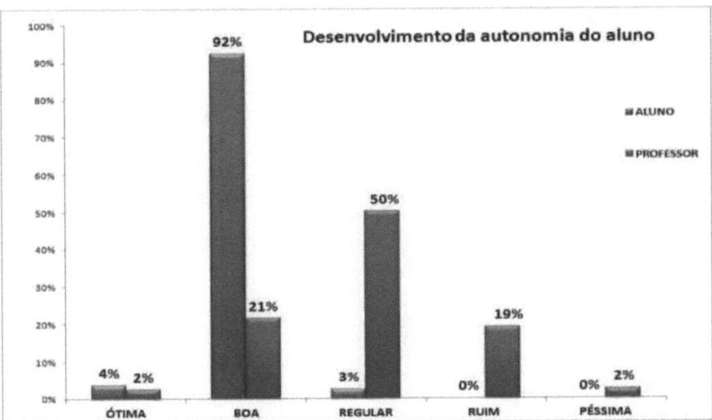

Graph 13 - Development of student autonomy (Organization) Source: survey data.

According to Graph 13, 92% of the respondents consider the development of their study autonomy to be GOOD during the period in which they were taking semipresential courses. While 50% consider this development REGULAR.

It is known that Heutagogy is a directed study, a self-learning process based on practical experiences and that, once in a safe environment, the more you make mistakes, the more you learn. Through technology, students can not only define "how", but also "when and where to learn" Batista (2008, p. 32). In this model, the student, who is responsible for their own development, defines when, how and what to learn. It was found that the VLE at Faculdades Vale do Carangola and the teachers are not aware of this learning model, and the students' judgment was based on their own concepts, instinctively.

The explanation for this is that the teachers who work in this teaching system do not carry out a survey of the student's profile, nor do they advocate certain learning theories for the preparation of online teaching material.

When a teacher was asked about the subject, the following response was obtained:

Teacher: *First of all, I'm not familiar with the term heutagogy, so I wouldn't be able to apply it in my classes.*

Teacher (b): I *knew the term, but not its concepts. But I find it interesting and I could certainly apply it in my classes*

Teacher (c): *That would be a lot of work. Since I'd have to put the same content in several different ways for the student to choose from.*

It is suggested that an environment conducive to the development of Heutagogy would be one rich in media, with mechanisms designed for different types of learning. Teachers should, first and foremost, propose different didactics to cater for different types of learning.

In this way, the student would qualify as a learner who ventures into their learning without the presence of the teacher or facilitator, since they would be doing something out of their experiences acquired throughout their life and according to their learning model.

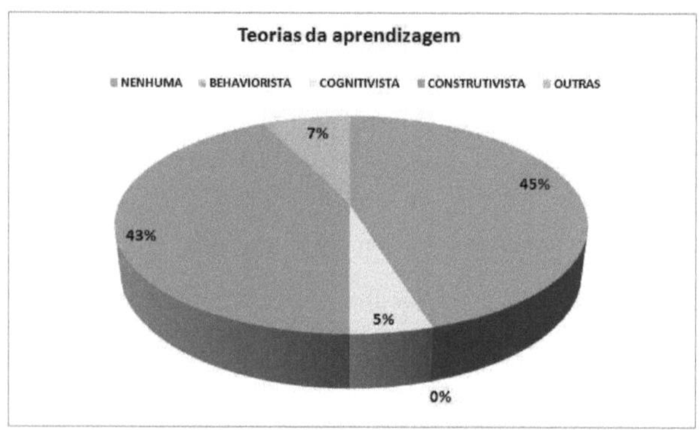

Graph 14 - Learning theories used by teachers in the production of teaching material for semi-presential courses.

Source: research data.

Graph 14 shows that 45% of the teachers interviewed do not use any learning theory as a parameter for producing *online* teaching materials. 43% produce teaching materials based on the Constructivist theory, 5% on the Cognitivist theory and 7% use other learning theories. No teacher uses the Behaviorist theory.

In the specific case of Faculdades Vale do Carangola, teachers are not used to applying the models indicated by learning theories in the production of *online* teaching material in the development of semi-presential courses. It is believed that this is due to the practicality and direction of management, which directs course activities through standard behavioral guidelines. In this document, management asks teachers to post in the virtual environment: two discussion forums, two videos and two activities on the subject being taught.

The use of this type of resource in distance education is based on a number of theories. Among these is Ausubel's (2002) Theory of Significant Learning, which postulates that learning takes place through the assimilation of new concepts and propositions and depends on a pre-existing conceptual framework, i.e. propositional systems already possessed by the learner.

According to Siemens (2004), the theories most used in the creation of teaching materials are Behaviorist, Cognitivist and Constructivist. But this conception, dating from 2004, does not prevail in the context of this turn of the century.

Graph 15 - Theory underpinning the production of teaching materials Source: survey data.

According to Graph 15, 55% of the teachers who use some kind of learning theory use theories to support the production of *online* teaching materials.

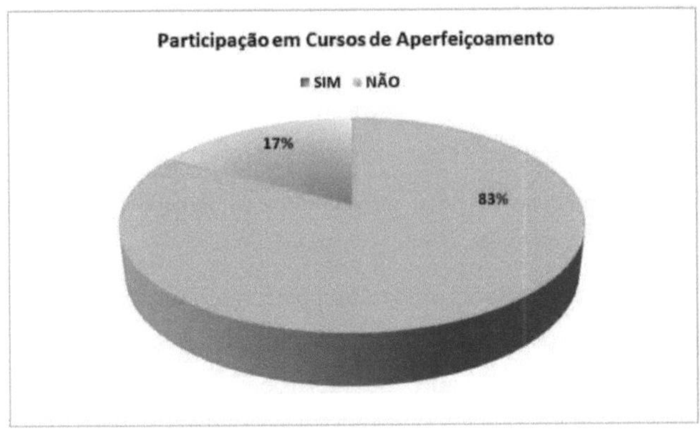

Graph 16 - Training or specialization course taken by the teacher in the production of online teaching material Source: survey data.

Graph 16 shows that 83% of the teachers who teach semi-presential subjects have taken part in EaD training courses to produce *online teaching* materials.

It is important to note that more and more teachers are being exposed to the use of distance or semi-distance learning, either for their own training or as the person responsible for subjects in this modality. This makes it necessary to train them to produce material or to take on the role of the new teacher in the face of new technologies.

It should be emphasized that pedagogical practice requires teachers to be trained from a multi-referential, political, technical and human perspective, since it articulates plural knowledge - pedagogical, experiential, scientific, technological and political - in a sense of engagement with social reality (FERREIRA; COELHO, 2006).

Thus, it can be seen that teachers who practice teaching *online* in VLEs must stay connected in their continuing education, however, it seems strange that many teachers still neglect the distance learning and semi-presential modalities.

A lot has been done to qualify teachers in line with the demand offered by the communication unleashed by digital technologies in the search for quality teaching.

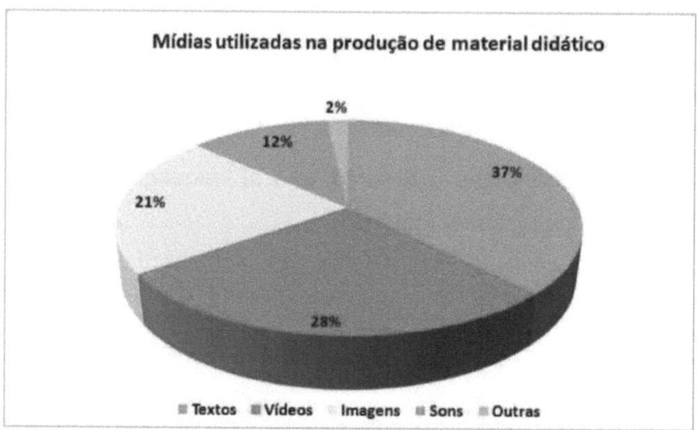

Graph 17 - Media used by teachers to produce teaching material for semi-presential courses
Source: survey data.

In Graph 17, the media most used by teachers as *online* teaching material is text, with 37% of respondents; video with 28%; images with 22% and 12% using sounds. This analysis is confirmed in Figures 8 and 9, which show the classrooms of three teachers who teach semi-presential courses.

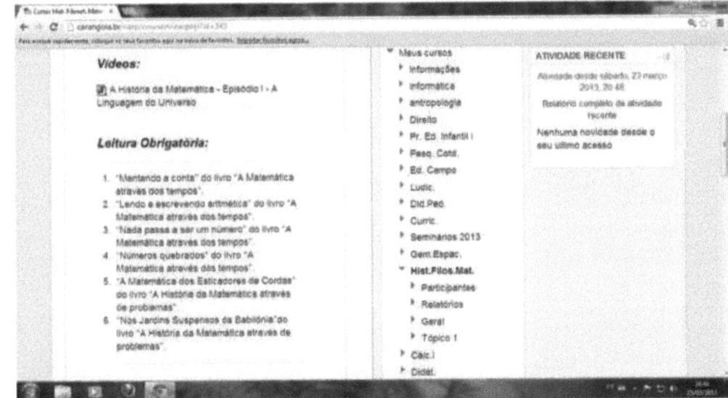

Figure 8 - History and philosophy of mathematics virtual classroom Available at: <http://.carangola.br/casp>. Accessed on: March 25, 2013.

Figure 8 confirms the high incidence of the text-to-read tool in the virtual learning environment.

What is striking, however, is that authors with a command of knowledge on the subject are invoked. The institution does not have an interdisciplinary team for its own production, as Effting (2010) suggests, where each teacher is responsible for a portion of the production and results in joint work.

Among the current resources that are explicit in Ava Moodle, Silva (1998), suggests using hypertext/hypermedia, because this type of text

[...] allows interactive linking of information, in a non-linear way, presented in different formats that can include text, static or animated graphics, film clips, sounds and music. It also provides flexible means for users to organize and manage information in order to satisfy different needs and to be able to assemble and produce complex, large, richly connected and cross-referenced collections (p. 12).

Santaella (2001) points out that while printed text is characterized by a linear flow, hypertext breaks down this linearity into smaller units or blocks of information, the basic attributes of which are nodes and associative media, forming a system of connections that make it possible to link one node to another by means of hyperlinks.

Therefore, links work as a shortcut or as a projection movement between the limits of what has been read and other topics related to the same subject, complementing, reaffirming or even contradicting it.

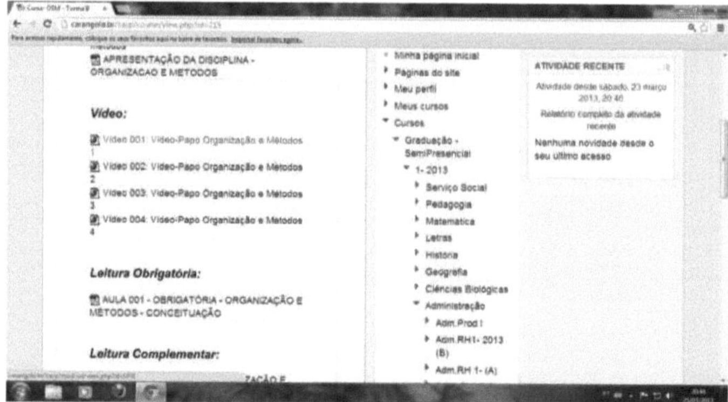

Figure 9 - Virtual classroom for a course in Administration Available at: <http://.carangola.br/casp>. Accessed on: March 25, 2013.

Figure 9 shows that there is a large concentration of videos posted and a few texts. Moran (2000, p. 38) defends the use of video from various perspectives. According to the researcher, television and video combine sensory-synesthetic information with audiovisual information and communicate with the majority of people, both children and adults. He also states that audiovisual messages require little effort and involvement on the part of the recipient.

Audiovisual language. It develops multiple perspectives: it constantly solicits imagination and reinvests affectivity with the role of primary mediation in the world, while written language develops more rigor, organization, abstraction and logical analysis (idem, p. 39).

Santaella (2008), when explaining the world of images, goes further when he says that our visual representations are part of a world that is in the realm of the perceptual and classifies some media as: paintings, engravings, photographs and cinematographic, television, videographic and infographic images.

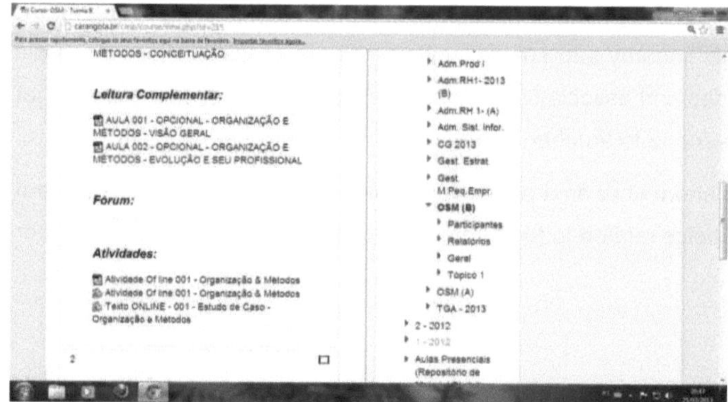

Figure 10 - Virtual classroom of the subject Organization and methods of the Administration course

Available at: <http://.carangola.br/casp>. Accessed on: March 25, 2013.

Figure 10 shows that the activities are concentrated on producing and reading texts.

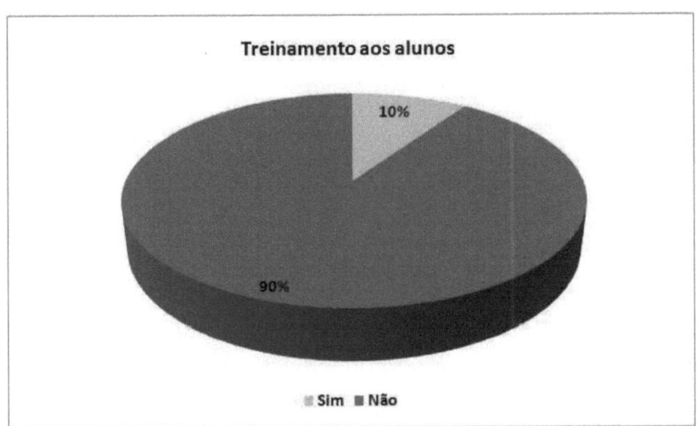

Graph 18 - Training given to students by teachers before the semi-presential course was taught
Source: survey data.

It can be concluded from Graph 18 that 90% of the interviewees do not train their students at the start of the semi-presential courses and 10% train their students before they start the semi-presential course.

In order to set up a learning and training environment, Filatro recommends creating a multidisciplinary team to develop teaching content, made up of teachers, tutors and support staff. Only then should training courses be created for both students and teachers using AVA Moodle.

According to Silva (2011, p. 12), Moodle:

It has content interfaces capable of creating, managing, organizing and moving a complete documentation (texts, graphics, images, videos, audios) and communication interfaces capable of promoting authorship and collaboration (e-mail, forum, chat, wiki, blog).

It is believed that the VLE training given to students before the start of a semi-presential course can alter the flow of classes in terms of time. With the appropriation of this knowledge, more specific than technical questions can be raised, without jeopardizing the time cycle planned for the course.

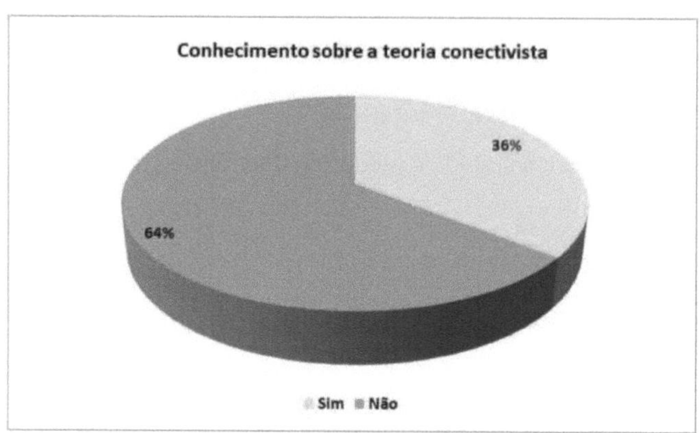

Graph 19 - Knowledge of Connectivist Learning Theory Source: survey data.

Graph 19 shows that 64% of teachers don't know about connectivist learning theory and 36% are aware of it, but haven't put it into practice or appropriated it yet. It is assumed that this is due to the lack of research into this theory and its correlates. As Siemens (2004) himself informs us, this is a theory for the digital age, and the majority of teachers working in the education system were born before this age, so they have no contact with the relatively new theory.

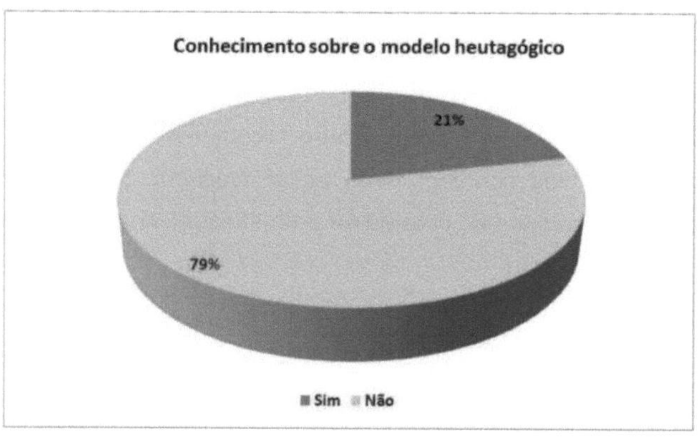

Graph 20 - Knowledge of the self-directed education model (heutagogy) Source: survey data.

Graph 20 confirms that 79% of the teachers surveyed are not familiar with the heutagogical education model. It is believed that this lack of knowledge is due to the fact that Heutagogy is also a relatively new science.

As Peters (2003) points out, models based on self-learning have certain implications. The self-directed learning movement propagated by Heutagogy should be viewed with caution, as it indicates what can or should be learned on one's own, with the student taking full responsibility for their learning process.

In this way, attention should be paid to educational policies that are implicit in ideas of autonomy, as indicated by Moura (1997). Heutagogy's proposal is to indicate mechanisms that support the provision of distance learning or bimodal courses, based solely on interactivity and participation, with intense use of instructional material.

This proposal, based on self-learning, calls for the prior planning of subjects, and as Belan (2008) explains, it is necessary to know the profile of the learner in this teaching modality, thus outlining their characteristics, needs and experiences.

6   CONCLUSION

The exercise undertaken in this research covered some of the dimensions that make up the theoretical basis of the research object and, to this end, we sought to analyze some aspects of Connectivist Learning Theory, which should not be left out of reflection; resulting in research that produced many developments and demonstrated integration between fields of study.

In the course of the research, an attempt was made to find a way of applying connectivist theory to the traditional instructional design of semi-presential courses at the Faculdades Vale do Carangola. This proposal arose from the need for theoretical support in the production of *online teaching* materials for the semi-presential modality at this higher education institution. It was found that learning theories were not being used in the production of teaching materials.

It was necessary to look at other themes such as networked knowledge, andragogy and heutagogy. All of them have theoretical traits that promote a connection between their particularities.

Next, it was found that the connectivist theory, defined as an alternative theory of the digital age and revealed by Siemens (2004), presupposes that learning is internalized in the individual and that it needs to be triggered by a source of knowledge that can reside both in other individuals and in non-human devices. Hence the need to make and create connections to trigger internalized knowledge.

According to the Connectivist Theory, learning can occur through connections with people, electronic equipment or databases, and the ability to recognize this connectivity is what sets the individual up for meaningful learning.

Semi-presential teaching, which is a hybrid form of education, combining face-to-face and distance modalities, and the lack of theoretical support experienced in the research for the production of didactic material, provided a confrontation with other variants.

From the realization that there was a need for more comprehensive research that could support or redirect the paths that structure the semi-presential teaching modality and the emerging methods from information and communication technologies, the research was outlined.

However, none of the references point to a specific formula for teachers to conduct a semi-presential course, given that students have different habits, backgrounds, cultures and experiences. However, knowing the profile of the students they will be working with is important for the teacher to be able to conduct their *online* activities satisfactorily, adapting the VLE to the needs and characteristics of each student.

Another important factor observed is that the teacher is tied to the face-to-face content, which has few technological tools for its application, and is the center of the learning process. This systemic vision of the teacher as a mediating agent must be implemented together with a holistic vision in which each part of the subject must be considered and parameterized in accordance with the planning.

This planning should include the timetable for the subject and each member of the support team - class council and teacher - should be able to give a democratic opinion on the implementation.

In this way, the teacher must conduct his or her subject, first by surveying the profile of the class and through this research, plan the content by selecting the tools for its application. These choices should promote debate and provide up-to-date information to help students build knowledge.

In the knowledge society, students are connected and communication tools can strengthen learning through knowledge, social and digital networks. These factors, advocated by Connectivism as being of great value for learning to take place, can be observed, tested and used in VLEs for teaching subjects in the semi-presential modality.

In carrying out the survey, which described the quality of semi-presential teaching at the Vale do Carangola Faculties, it was observed that the responses obtained, taking into account the input aspects, are an analysis of resources, technology and teacher training. There was still a need for teacher training and investment in technology for it to be considered "good" quality.

As for the technological resources available, Faculdades Vale do Carangola has a 10 megabit Internet access link, a computer laboratory with 20 microcomputers, a service and support sector with only one employee to meet the demand of all the face-to-face and semi-presential courses. Therefore, the service provided to specific students in the hybrid modality is compromised. For this reason, it is understood that the quality benchmark has not yet been achieved.

With regard to the process aspects, which look at the pedagogical and organizational context and relate to direct participation in the education process, the question of quality was repeated, when the data survey indicated -regular and good" quality. This result was blamed on the fact that the majority of students considered planning to be -regular". In reality, planning is done along the same lines as in-class courses, but with the application of technology.

As aspects of the results process that analyze the characteristics of the intermediate and final purposes of education, it was found that the evaluations carried out by the VLE in the hybrid and face-to-face modality always have predominantly face-to-face characteristics.

It is believed that in the hybrid modality, diagnostic, summative and formative evaluations would fit, linked only by the degree of connection between the students and the subjects and the tools used by the teachers.

In view of this, it was concluded that the success of distance learning students is directly linked to a number of factors, and that both the connectivist learning theory and the andragogical and heutagogical principles can be associated with learning in the hybrid modality; however, they still constitute totally new areas of study and research, which offer scope for unpublished studies for the development of research into distance learning.

It is therefore believed that the Connectivist Theory as a theoretical model underpinning the production of didactic materials for Virtual Teaching Environments in the hybrid modality can fill in some of the probable gaps left by other theories and further favor the training of both the e-student and the e-teacher, and that this process takes place through participation, interaction and cooperation promoted by communication in information and knowledge networks.

BIBLIOGRAPHICAL REFERENCES

ASSIS, Simone Ferreira de. **The evaluation criteria used by the tutor of the discipline theories of administration in the chat, forum, wiki and questionnaire tools**. SIED-International Symposium on Distance Education. Enpd. Meeting of Researchers in Distance Education. Federal University of Sao Carlos (UFSCAR). September, 2012.

AUSUBEL, David Paul; NOVAK, Joseph D.; HANESIAN, Helen. **Educational Psychology**. Rio de Janeiro: Interamericana, 1980.

AUSUBEL, David Paul. **Psicologia educativa:** un punto de vista cognoscitivo. Mexico City: Trillas, 1976.

AUSUBEL, David Paul. **Acquisition and retention of knowledge.** A cognitive perspective. Barcelona, Paidós, 2002.

ALMEIDA. M. E. B. **Educaçâo a distância na internet:** abordagens e contributes dos ambientes digitais de aprendizagem. Educaçâo e Pesquisa, Sâo Paulo, v. 29, n. 2, p.327-340, dez. 2003. Available at: <http:www.scielo.br/pdf/ep/v29n2/a10v29n2.pdf>. Accessed on: Feb. 24, 2013.

ALTOÉ, Anair; PENATI, Marisa Morales. Constructivism and constructionism based on teaching action in a computerized environment. In: ALTOÉ, Anair; COSTA, Maria Luisa Furlan; TERUYA, Teresa Kazuko (Org.). **Educaçâo e novas tecnologias.** Maringà: Eduem, 2005. p.55-68

BERNHEIM, Carlos Tünnermann; CHAUi, Marilena de Souza. **Challenges of the university in the knowledge society:** five years after the world conference on higher education. Brasilia: UNESCO, 2008. 44 p.

BAPTISTA, Maria da Nazaré Mesquita Martins dos Santos. **Poiésis,** Tubarâo, v. 4, n. 7, p. 145-155, jan./jun. 2011.

BARNES, J. A. **Social Networks**. Cambridge: Module 26, p. 1-29, 1972.

BARROS, C. S. G. (1998). **Points of School Psychology**. Sâo Paulo: Editora Atica.

BEHRENS, Marilda Aparecida; OLIARI, Anadir Luiza Thomé. The evolution of paradigms in education: from traditional scientific thinking to complexity. **Diàlogo Educ.**, Curitiba, v. 7, n. 22, p. 53-66, Sept./Dec. 2007.

BELLAN, Zezina. Heutagogy - **Learn to Learn More and Better**. Santa Barba d'Oeste: SOCEP Editora, 2008.

BELLAN, Zezina. **Andragogy in Action:** how to teach without becoming boring. Editora Z3, 2005, 156 p.

BOCK, Ana; FURTADO, Odair; TEIXEIRA, Maria. **Psychologies. An introduction to the study of Psychology**. Sao Paulo: Saraiva, 1999. p. 38-47.

BOYLE, Tom. 1997. **Design for Multimedia Learning**. London: Prentice Hall.

BIAGGIO, Angela Maria Brasil. **Developmental psychology**. 4. ed. Petrópolis: Vozes, 1976.

BRAZIL. Ministry of Education. **Ordinance 4.059 of 2004**. Available at: <http://portal.mec.gov.br/sesu/arquivos/pdf/nova/acs_portaria4059.pdf>. Accessed on: 05 Nov. 2012.

BRAZIL. National Education Council. **CNE/CP Resolution No. 1/2002, of February 18, 2002**. Establishes National Curriculum Guidelines for the training of Basic Education teachers, at higher education level, full degree course. Brasilia, 2002. Available at: <http://www.mec.gov.br/cne>. Accessed on: 05 Nov. 2012.

BRITO, A. E. Training teachers: rediscussing work and teaching knowledge.

In: MENDES SOBRINHO, J. A. de C. (Org.). **Formação de professores e práticas docentes: olhares contemporâneos**. Belo Horizonte: Autêntica, 2006.

BEHRENS, Marilda Aparecida. Collaborative learning projects in an emerging paradigm. In: **New Technologies and Pedagogical Mediation.**

Campinas, SP: Papirus, 2000. p. 67-132

BELLAN, Zezina. Heutagogy - **Learn to Learn More and Better**. Santa Barba d'Oeste: SOCEP Editora, 2008.

BELLONI, Maria Luisa. **Distance Education**. Campinas: Autores Associados, 2006.

BERTOLIN, J. C. G. **Avaliaçao da qualidade do sistema de educação superior brasileiro em tempos de mercantilizaçao - Periodo 1994-2003**. 2007. Thesis (Doctorate in Education) - Federal University of Rio Grande do Sul, Porto Alegre, 2007.

BORGES, Martha Kaschny. **Semi-presential education:** demystifying distance education. Available at: <http://www.abed.org.br/congresso2005/por/pdf/218tcf3.pdf>. Accessed on: August 9, 2012.

CASTELLS, Manuel. **The network society**. Sâo Paulo: Paz e Terra, 1999.

CARELLI, Rogério. Historian Professor at Vale do Carangola College. **Interview** conducted on July 28, 2008.

CARNEIRO, D. V.; COSTA JUNIOR, J. M.; NUNES, V. B.; NOBRE, I. A. M.; BALDO, Y. P. **A planning proposal for creating rooms in the Moodle virtual learning environment:** adapted activity map. In: XVI

ABED International Congress on Distance Education, Foz do Iguaçu - PR, 2010.

COSTA, Julio Resende. Analysis of the Instructional Design of the Teacher Training Course in Youth and Adult Education. **Interscience Place,** International Scientific Journal, v. 1, n. 6, Apr./Jun. 2012.

CREECH, Heather; WILLARD, Terri. **Strategic intentions:** managing knowledge networks for sustainable development. Winnipeg: IISD - International Institute for Sustainable Development, 2001. Available at: <http://www.iisd.org/pdf/2001/networks_strategic_intentions.pdf>. Accessed on: Dec. 21, 2012.

CUNHA, Luisa Margarida Antunes da. **Rasch models and Likert and Thurstone scales in the measurement of attitudes**. Master's dissertation. University of Lisbon, Faculty of Sciences. Department of Statistics and Operational Research, 2007.

DEMO, Pedro. Education today: **"new" technologies, pressures and opportunities**. Sâo Paulo: Atlas, 2009

EYSENCK, M. W.; KEANE, M. T. **Cognitive Psychology.** Porto Alegre: ARTMED, 1994.

EBERT, Cristiane do Rocio Cardoso. Semi-presential teaching as a response to the growing need for permanent education. **Educar em Revista,** n. 21, p. 1-16, 2003.

FETTERMANN, Joyce Vieira. **The virtual environments of the My English Club social network and their interventions in face-to-face English language learning environments**. 2012. Dissertation (Master's Degree in Cognition and Language) - Postgraduate Program in Cognition and Language, Universidade Estadual do Norte Fluminense, Campos dos Goytacazes, RJ, 2012.

FERREIRA, Maria da Conceiçâo Alves; COELHO, Maria das Graças Pinto. **Online teaching:** weaving possibilities into educational practice. Esud, 2006.

FILATRO, Andrea. **Contextualized instructional design**: education and technology. Sâo Paulo: Editora SENAC, Sâo Paulo, 2004.

FILATRO, Andrea. Instructional Design in practice. **Person Education do Brasil,** 2008. 173 p.

FONSECA, Vitor da. **Introduçâo às Dificuldades de Aprendizagem**. 2. ed. Porto Alegre: Artes Médicas, 1995.

FRANCO, Lùcia R. H.; BRAGA, Dilma B. Virtual Communication. Digital Book. **Instructional Design Course for Virtual Distance Education.** Itajubà: UNIFEI, 2007.

FRANCO, Augusto de. **Networks are environments for interaction, not participation.**
2010. Available at: <http://escoladeredes.net/ profiles/blogs/redes-sao-ambientes- de>. Accessed on: Jan. 21, 2013.

GARDNER, Howard. **Structures of the Mind: The Theory of Multiple Intelligences**. Porto Alegre: Artes Médicas, c1994. Originally published in English under the title: The frams of the mind: the Theory of Multiple Intelligences, in 1983.

GOMES, Maria de Fàtima Cardoso; SENA, Maria das Graças de Castro (Orgs.).
Learning difficulties in literacy. 2. ed. Belo Horizonte: Autêntica, 2004.

GOUVEIA, Sylvia Figueiredo. The paths of teachers in the age of technology. **Revista da educaçâo e informàtica**, year 9, n. 13, apr. 1999.

GUTIERREZ, Francisco; PRIETO, Daniel. **Pedagogical mediation: alternative distance education**. Campinas: Papirus, 1994.

GRENNE, J. **Thought and Language**. Rio de Janeiro: Zahar, 1976.

LÉVY, P. **The Technologies of Intelligence:** the future of computer thinking. Rio de Janeiro: Ed34, 1993. 203p.

LÉVY, Pierre. **Collective intelligence:** towards an anthropology of cyberspace. Sao Paulo: Loyola, 1998.

LÉVY, Pierre. **What is the virtual?** Rio de Janeiro: Ed. 34, 1996.

LÉVY, Pierre. **Cyberculture**. Sao Paulo, Ed. 34, 1999.

LÉVY, Pierre. **The technologies of intelligence:** the future of thought in the computer age. Translated by Carlos Irineu da Costa. Rio de Janeiro: Editora 34, 2006.

LITWIN, Edith (Org.) **Educational technology**. Politics, histories and proposals. Porto Alegre: Artes Médicas, 1997.

LOPES, Joice. **Social Networks**. Available at : <http://digartmedia.wordpress.com/ 2010/06/03/redes-sociais-3/>. Accessed on: November 29, 2011.

KAUARK, Fabiana; MANHÂES, Fernanda Castro; SOUZA, Carlos Henrique Medeiros de.

Metodologia da pesquisa: guia pràtico. Itabuna: Via Litterarum, 2010. 88p.

KNOWLES, Malcolm S. **Andragogue versus pedagogue**. Association Press, USA, 1980.

KENSKI, Vani M. **Tecnologias do Ensino Presencial e a Distância**. Campinas, SP Papirus, 2003 - (series pràtica pedagògica).

KENSKI, Vani M. Novas tecnologias: o redimensionamento do espaço e do tempo e os impactos no trabalho docente. **Revista Brasileira de Educaçâo**, n. 8, p. 58-71, May/Aug. 1998.

KERR, Bill. **A Challenge to Connectivism**. Transcript of paper presented at the *Online Connectivism Conference*, February 2007, University of Manitoba. Available at

at: <http://ltc.umanitoba.ca/wiki/index.php?title=Kerr_Presentation>. Accessed on: 21 Jan. 2012.

KOP, Rita; HILL, Adrian. Connectivism: Learning theory of the future or vestige of the past? *The International Review of Research in Open and Distance Learning*, v. 9, n. 3, 2008. Available at: <http://www.irrodl.org/index.php/irrodl/article /view/523/1103>. Accessed on: January 31, 2012.

HAYDT, R. C. C. **Avaliaçâo do processo ensino-aprendizagem**. Sao Paulo: Atica, 1988.

HARTREE, A. Malcolm Knowles' theory of andragogy: a critique. **International Journal of Lifelong Education**, v. 3, 1984.

HASE, Stewart; KENYON, Chris. **From Andragogy to Heutagogy**. UltiBase, December 2000. Available at: <http://ultibase.rmit.edu.au/Articles/dec00/ hase1.pdf>. Accessed on: Jan. 21, 2012.

HEIDE, Ann; STILBORNE, Linda. **Teacher's guide to the Internet:** complete and easy. 2. ed. Porto Alegre: Artes Médicas Sul, 2000.

HOFFMANN, J. M. L. **Avaliaçâo: mito e desafio:** uma perspectiva construtivista. Porto Alegre: Mediaçao, 2001.

LANDIM, Claudia Maria das Mercês Paes Ferreira. **Educaçâo à Distância: algumas considerçôes.** Rio de Janeiro: Vozes, 1997.

LIKERT, R. Una Técnica para la Medicion de Atitudes (A technique for the measurement of attitudes, Archives of Psychology, n.140, p.1-50, 1932). In: WEINERMAN, C. H. **Escalas de Medicion en Ciências Sociales.** Buenos Aires: Nueva Vision, 1976. p. 201-

260.

LITWIN, Edith (Org.). **Educaçâo a Distância:** Temas para Debate de uma Nova Agenda Educativa. Porto Alegre, Artmed, 2001.

LÓPEZ SEGRERA, F. **Globalization and higher education in Latin America.**

Caracas: UNESCO-IESALC, 2001.

MARCONI, Marina de Andrade; LAKATOS, Eva Maria. **Metodologia do trabalho cientifico.** 4. ed. Sao Paulo: Editora Atlas, 1992. p. 43-44.

MATTAR, Fauze Najib. **Marketing research.** 3. ed. Sao Paulo: Atlas, 2006. 224 p.

MINISTRY OF EDUCATION - MEC. **Quality Benchmarks for Distance Higher Education** Distance Education Secretariat. Brasilia, August 2007. Available at: http://portal.mec.gov.br/seed/arquivos/pdf/legislacao/refead1 .pdf.

Accessed on: March 21, 2013.

MINISTRY OF EDUCATION - MEC. **PORTARIA N° 4.059, DE 10 DEZEMBRO DE 2004** (DOU de 13/12/2004, Seçâo 1, p. 34). Available at: <http://portal.mec.gov.br/sesu/arquivos/pdf/ nova/acs_portaria4059.pdf>. Accessed on: October 23, 2012.

MORAN, José Manuel. **Novas Tecnologias e mediação pedagógica** - Campinas, SP: Papirus, 2000 (Coleçâo Papirus Educaçâo).

MOREIRA, M. A. David Ausubel's learning theory. **Fasciculos do CIEF**, Teaching-Learning Series, n. 1, 1993.

MOREIRA, Marco Antônio. **Theories of Learning**. Sâo Paulo: Epu, 1999. 195 p.

MORAN, José Manuel. **Proposals for changes in face-to-face courses with online education**. Available at: <http://www.eca.usp.br/prof/moran/propostas. htm#novos>. Accessed on: June 21, 2012.

MORAN, José Manuel. **The education we want:** new challenges and how to get there. Papirus Publishing House. Campinas - SP. 2007. 3. ed. 173 p.

MORAES, Maria Cândida. **The emerging educational paradigm**. Campinas: Papirus, 1997.

MOREIRA, Marco Antonio. **The theory of meaningful learning and its implementation in the classroom**. Brasilia: Editora UnB, 2006.

MOURA, R. M. **The process of self-directed learning in adults**. Unpublished master's thesis, Portuguese Catholic University, Faculty of Human Sciences, Lisbon, 1997. Available at: <http://rmoura.tripod.com/indicetese.htm>. Accessed on: January 25, 2013.

MORAN, José Manoel. **The new spaces in which teachers work with technologies**. Revista Diàlogo Educacional, Curitiba, PUC-PR, v. 4, n. 12, p.13-21, May/Aug. 2004. Available at: <http://www.eca.usp.br/prof/moran /espacos.htm>. Accessed on: October 12, 2012.

MORAN, José Manuel. **Proposals for changes in face-to-face courses with online education**. Available at: <www.ece.usp.br;prof/moran>. Accessed on: June 21, 2005.

MORAN, José Manuel. **New Technologies and Pedagogical Mediation** - Campinas, SP: Papirus, 2000 (Coleçâo Papirus Educaçâo)

MORAN, José Manoel. **New course models.** Paper presented at the 11th International Congress on Distance Education. 8/2009 in Salvador, BA. Available at: <http://www.eca.usp.br/prof/moran/propostas.htm #novos>. Accessed on: Feb. 14, 2013.

MOTA, José. **Connectivism and Learning on the Net** (Chap. IV). From Web 2.0 to e-Learning 2.0: Learning on the Net. 2009. Available at: <http://orfeu.org/weblearning20/4_2_conectivismo>. Accessed on: November 28, 2011.

MUGNOL, Marcio. Distance education in Brazil: concepts and foundations. **Rev. Diàlogo Educ.**, n. 27, p. 335-349, May/Aug. 2009.

NISKIER, Arnaldo. **Distance Education:** The Technology of Hope. São Paulo, Loyola, 1999.

OLIVEIRA, Ari Batista de. **The Essence of Andragogy for Business**. MEd Education. University of Minnesota - USA. Instituto Andragógico de Desenvolvimento Humano EDITORA iand. 2011

OLIVEIRA, Vera Barros. **Informatica em Psicopedagia**. Sâo Paulo: Editora SENAC, 1996.

OLIVEIRA, Sheila da Costa. Face-to-face meetings: an ODL tool? **New Technologies in Education,** v. 5, n. 2, dec. 2007.

PENNA, A. G. **Introduction to cognitive psychology**. Sâo Paulo. EPU, 1984.

PETERS, Otto. The **Didactics of Distance Learning:** Experiences and the State of the Discussion from an International Perspective. Trad. I. Kayser. S. Leopoldo/RS, Editora Unisinos, 2003

PERRENOUD, Philippe. **Ten New Skills for Teaching**. Porto Alegre: Artes Médicas Sul, 2000.

PIAGET, Jean. **The Psychology of the Child**. Porto: Asa, 1997.

PIAGET, Jean. **Where is Education Going?** Lisbon: Livros Horizonte, 1990.

PORTAL, Leda Lisia Franciosi. Distance education: a strategic-methodological option in search of spaces of distance or relationship for learning. **Rev. Educaçâo**, n. 44, p. 93-115, Aug. 2001.

PRETI, O. Educaçâo a distância: uma pràtica educativa mediadora e mediatizada. In: PRETI, O. (Org.). **Educaçâo a distância:** inicio e indicios de um percurso. Cuiabà: UFMT, 1996.

QUEVEDO, Angelita. Semi-presential teaching from the student's point of view. **Revista e-curriculum,** v. 7, n.1, abr. 2011. Available at: <http://revistas.pucsp.br/index.php/curriculum/article/view/5678>. Accessed on: October 31, 2012.

REZENDE, Denis Alcides; ABREU, Aline França. **Information Technology Applied to Business Information Systems**. Sao Paulo: Ed. Atlas, 2000.

REZENDE, Flàvia Amaral. **Characteristics of the Constructionist virtual teaching and learning environment in the training of university teachers**. 2004. 246 f. Dissertation (Master's Degree in Multimedia) - Institute of Arts. State University of Campinas, Campinas, 2004. Available at: <http://www.iesalc.unesco.org.ve/documentosinteres/brasil/caracteristicas%20 del %20aprendizaje%20virtual %20-%20Flavia%20Amaral%20Lazende.pdf>. Accessed on: 21 Mar. 2013.

SANTAELLA, L. **Matrizes da linguagem e do pensamento:** sonora, visual, verbal. Sao Paulo: Iluminuras, 2001.

SANTAELLA, Lucia; NOTH, Winfried. **Image**: cognition, semiotics, media. Sao Paulo: Iluminuras, 2008.

SKINNER, B. F. **On behaviorism**. Sao Paulo: Cultrix (1974).

SOUZA, Carlos Henrique Medeiros de; GOMES, Maria Lucia Moreira. **Education and Cyberspace**. Brasilia: Editora Usina de Letras, 2008. 158 p.

SIEMENS, George. **Connectivism:** A Learning Theory for the Digital Age. 2004. Available at: http://www.elearnspace.org/Articles/connectivism.htm>. Accessed on: Nov. 28, 2011.

SIEMENS, George. **Learning Development Cycle:** Briding Learning Design and Modern Knowledge Needs. July, 2005.

SIEMENS, George. *^Qué tiene de original el conectivismo?* 2008. Available at: <http://humanismoyconectividad.wordpress.com/2009/01/14/conectivismo-   siemens/>. Accessed on: October 26, 2012.

SILVA, Robson Santos da. **Moodle para autores e tutores**. 2. ed. Sao Paulo: Novatec, 2011.

SOARES, Magda. **Literacy and literacy**. Sao Paulo: Contexto, 2002.

STAATS, A. W. Social behaviorism: a science of man with freedom and dignity. **Arquivos brasileiros de psicologia,** v. 32, n. 4, p. 97-116, 1980.

STERNBERG, R. J. **Cognitive Psychology**. Porto Alegre: ARTMED, 2000.

TENNANT, M. **Psychology and adult learning**. 2. ed. London: Routledge, 1997. 160 p.

TOFFLER, Alvin. **Powershift**. Rio de Janeiro: Record, 1990.

TORI, Romero. Hybrid courses or blended learning. In: LITTO, Frederic M.;

FORMIGA, Marcos (Orgs.). **Distance education:** the state of the art. Sao Paulo: Pearson Education, 2008. p. 121-128.

TURKLE, Sherry. **Life on the screen** - identity in the internet age. Lisbon, Relógio D'Agua Ed., 1997.

TOMAÉL, Maria Inês. **Knowledge Networks.** Datagramazero (Rio de Janeiro), v. 9, p. 2, 2008. Available at: <http://www.dgz.org.br/abr08/Art_04.htm>. Accessed on: Dec. 21, 2012.

SA, Iranita M. A. **Educaçâo a Distância:** Processo Continuo de Inclusao Social. Fortaleza, C.E.C., 1998.

UNESCO - UNITED NATIONS EDUCATIONAL, SCIENTIFIC AND CULTURAL ORGANIZATION. **Conceptual framework**. Documentos Laboratorio Latinoamericano de Evaluación de la calidad de la educación. Santiago, Chile: Lecce, Orealc/Unesco, 1997.

VENTURA, P. C. S. La négociation commme élément d'interprétation dans la communication, la vulgarisation et l'éducation en science et technique. In: CONGRESO MUNDIAL DE CIENCIAS DE LA EDUCACIÓN, 14., 2004, Santiago. **Proceedings...** Santiago, 2004.

VERHAGEN, Plon. **Connectivism:** a new learning theory? 2006. Available at

<http://www.surfspace.nl/nl/ Redactieomgeving/ Publicaties/Documents/ Connectivism%20a%20new%20theory.pdf>. Accessed on: July 19, 2009.

VOGT, Maria Saleti Lock; ALVES, Elioenai Dornelles. Theoretical review of adult education for an approach to andragogy Education. **Revista do Centro de Educaçâo**, v. 30, n. 2, p. 195-213, jul./dic. 2005.

VOIGT, Emilio. The bridge over the abyss: semi-presential education as a challenge of the new times. **Revista Estudos teológicos**, v. 47, n. 2.2002. Available at: <http://www3.est.edu.br/publicacoes/estudos_teologicos/vol4702_ 2007/ET2007-2c_evoigt.pdf>. Accessed on: 06 Nov. 2012.

VOIGT, Emilio. The bridge over the abyss: semi-presential education as a challenge of the new times. **Revista Estudos teológicos**, v. 47, n. 2, 2002. Available at: <http://www3.est.edu.br/publicacoes/estudos_teologicos/ vol4702_2007/ET2007-2c_evoigt.pdf>. Accessed on: 06 Nov. 2012.

yes
I want morebooks!

Buy your books fast and straightforward online - at one of world's fastest growing online book stores! Environmentally sound due to Print-on-Demand technologies.

Buy your books online at
www.morebooks.shop

Kaufen Sie Ihre Bücher schnell und unkompliziert online – auf einer der am schnellsten wachsenden Buchhandelsplattformen weltweit! Dank Print-On-Demand umwelt- und ressourcenschonend produzi ert.

Bücher schneller online kaufen
www.morebooks.shop

Printed by Books on Demand GmbH, Norderstedt / Germany